MASTERING CYBER RESILIENCE

Second Edition

Kip Boyle

Jason Dion

Alyson Laderman

DISCLAIMER

While AKYLADE carefully ensures the accuracy and quality of these materials, we cannot guarantee their accuracy, and all materials are provided without any warranty whatsoever, including, but not limited to, the implied warranties of merchantability or fitness for a particular purpose. The name used in any data files provided with this course is that of a fictitious company and fictional employees. Any resemblance to current or future companies or employees is purely coincidental. If you believe we used your name or likeness accidentally, please notify us, and we will change the name in the next revision of the manuscript. AKYLADE is an independent provider of certification solutions for individuals, businesses, educational institutions, and government agencies. The use of screenshots, photographs of another entity's products, or another entity's product name or service in this book is for educational purposes only. No such use should be construed to imply sponsorship or endorsement of this book by nor any affiliation of such entity with AKYLADE. This book may contain links to sites on the Internet that are owned and operated by third parties (the "External Sites"). AKYLADE is not responsible for the availability of, or the content located on or through any External Site. Please contact AKYLADE if you have any concerns regarding such links or External Sites. Any screenshots used for illustrative purposes are the intellectual property of the original software owner.

TRADEMARK NOTICES

AKYLADE®, AKYLADE Certified Cyber Resilience Fundamentals®, A/CCRF®, AKYLADE Certified Cyber Resilience Practitioner®, and A/CCRP®, are registered trademarks of AKYLADE LLC in the United States and/or other countries. All other product and service names used may be common law or registered trademarks of their respective proprietors.

PIRACY NOTICES

This book conveys no rights in the software or other products about which it was written; all use or licensing of such software or other products is the responsibility of the user according to the terms and conditions of the software owner. Do not make illegal copies of books or software. If you believe that this book, related materials, or any other AKYLADE materials are being reproduced or transmitted without permission, please email us at report@akylade.com.

Copyright © AKYLADE LLC 2024

https://www.akylade.com

All rights reserved. Except as permitted under the United States Copyright Act of 1976, this publication, or any part thereof, may not be reproduced or transmitted in any form or by any means, electronic or mechanical, including photocopying, recording, storage in an information retrieval system, or otherwise, without express written permission of AKYLADE.

Paperback ISBN: 979-8-9886499-4-6

BONUS CONTENT

Please visit https://www.akylade.com/mastering-cyber-resilience to register your book and receive access to some online practice exams to help prepare you for your certification exams.

CONTENTS

Chapter 1 - Introduction 1
Chapter 2 - Cybersecurity Fundamentals 13
Chapter 3 - Risk Management Fundamentals 23
Chapter 4 - NIST Cybersecurity Framework 41
Chapter 5 - Framework Components 59
Chapter 6 - Six Functions 67
Chapter 7 - Controls and Outcomes 89
Chapter 8 - Implementation Tiers 111
Chapter 9 - Profiles 123
Chapter 10 - Assessing Cybersecurity Risk 151
Chapter 11 - The CR-MAP Process 165
Chapter 12 - Phase One: Discovering Top Cyber Risks 175
Chapter 13 - Phase Two: Creating a CR-MAP 207
Chapter 14 - Phase Three: Maintenance and Updates 231
Chapter 15 - Conclusion 241
Appendix A - A/CCRF Exam Objectives 245
Appendix B - A/CCRP Exam Objectives 257
Appendix C - Glossary 263

ACKNOWLEDGMENTS

After years of experience in the cybersecurity and certification industries, we discovered that there was a gap between earning certifications and the candidate's ability to perform specific job functions in the cyber resilience industry. Our goal is to help both job candidates and hiring managers bridge that gap by providing affordable and practice-based certifications that showcase your real-world skills.

A special thank you to AKYLADE's Advisory Council who is committed to this goal and who have provided invaluable insights, shared their expertise, and have dedicated countless hours in helping us develop these certifications.

This book is written for our community of students worldwide who have allowed us to continue to develop our courses and books over the years. Your hard work has led you to positions of increasing responsibility throughout your careers, and we are grateful to have been a small part of your success.

We truly hope that you all continue to love the method to our madness as you work to conquer version 2 of the AKYLADE Certified Cyber Resilience Fundamentals (A/CCRF) and AKYLADE Certified Cyber Resilience Practitioner (A/CCRP) certification exams.

We wish you all the best as you continue to accelerate your careers to new heights!

CHAPTER ONE

INTRODUCTION

In this book, you will learn how to master cyber resiliency in your organization as well as learn everything you need to know to pass version 2 of the AKYLADE Certified Cyber Resilience Fundamentals (A/CCRF) and AKYLADE Certified Cyber Resilience Practitioner (A/CCRP) certification exams. This book covers the essentials with no fluff, filler, or extra material, so you can easily learn the material and conquer the certification exams.

The AKYLADE Certified Cyber Resilience Fundamentals (A/CCRF) exam is the first certification exam in AKYLADE's Cyber Resilience certification path. This certification is designed to assess your theoretical understanding of the NIST Cybersecurity Framework version 2.0 (also known as NIST CSF or CSF herein) and your ability to plan, manage, and optimize its use within your organization.

The AKYLADE Certified Cyber Resilience Practitioner (A/CCRP) exam is the second certification exam in AKYLADE's Cyber

Resilience certification path. This certification is designed to test your ability in planning, managing, and optimizing the NIST Cybersecurity Framework version 2.0 within an organization by placing you into the role of a virtual cyber resiliency consultant throughout numerous different scenarios and real-world case studies.

This book assumes that you have no previous experience with the NIST Cybersecurity Framework version 2.0 and is designed to teach you exactly what you need to know to take and pass the AKYLADE Certified Cyber Resilience certification exams on your first attempt.

This text is designed to serve as a common body of knowledge for the certification exams and as a hands-on guide and workplace reference to operationalizing the NIST Cybersecurity Framework daily within your organization.

This book has also been designed to serve as the official textbook for the AKYLADE Certified Cyber Resilience series of certification exams. As such, this textbook has been divided into two portions.

The first portion of the book focuses on the basics of the NIST Cybersecurity Framework version 2.0. It is designed to aid in your studies for version 2 of the AKYLADE Certified Cyber Resilience Fundamentals (A/CCRF) certification exam. This portion consists of Chapter 2 through Chapter 10.

The second portion of the book focuses on the application of the NIST Cybersecurity Framework version 2.0 (CSF) using the proprietary Cyber Risk Management Action Plan (CR-MAP), which operationalizes and applies the CSF for use in the real world. If you are studying for the AKYLADE Certified Cyber Resilience Practitioner (A/CCRP) certification, you should focus on the second portion of the textbook; but be aware that everything covered in the first portion is fair game on this examination, too. This portion consists of Chapter 11 through Chapter 14.

Throughout the textbook, we will pause and visit various real-world organizations to observe how they have implemented the NIST

Cybersecurity Framework to increase cyber resilience within their organization. For each of these case studies, you gain insight into the successes achieved by various organizations across different industries and locations across the globe.

After the conclusion of the textbook, we have also included additional features to help you conquer the certification exams.

In Appendix A and Appendix B, you will find the exam objectives for each of the two certification exams, including the domains covered by each exam and their corresponding objectives.

To help you prepare for the AKYLADE Certified Cyber Resilience Fundamentals (A/CCRF) and AKYLADE Certified Cyber Resilience Practitioner (A/CCRP) certification exams, we have a full-length practice exam for each certification available for download at our website, https://www.akylade.com, which includes the practice exam, answer key, and explanations for each question.

If you fully understand the contents of this book and successfully complete the associated practice exam for the selected certification (scoring at least 85% or higher), you should be ready to take and pass your AKYLADE Certified Cyber Resilience Fundamentals (A/CCRF) and AKYLADE Certified Cyber Resilience Practitioner (A/CCRP) certification exams on your first attempt!

AKYLADE CERTIFIED CYBER RESILIENCE FUNDAMENTALS (A/CCRF)

The AKYLADE Certified Cyber Resilience Fundamentals (A/CCRF) certification exam is an entry-level certification used to demonstrate that a candidate understands all the material aspects of the NIST Cybersecurity Framework, including:

- The origin and original purpose of the framework

- The applicability of the framework across industries and sectors

- The three fundamental parts of the framework: the core, the implementation tiers, and the profiles

- The six functions: Govern, Identify, Protect, Detect, Respond, and Recover, as well as the 22 categories and 106 subcategories under the six functions

- The purpose, utility, and intended use of the Implementation Tiers, Profiles, and Informative References

- The use of the framework for identifying, assessing, and managing cybersecurity risk

Candidates who receive the AKYLADE Certified Cyber Resilience Fundamentals certification have demonstrated that they are ready to work as an entry-level cybersecurity specialist or consultant as part of a larger cybersecurity consulting team lead by an AKYLADE Certified Cyber Resilience Practitioner.

The AKYLADE Certified Cyber Resilience Fundamentals certification exam consists of 50 multiple-choice questions (40 graded questions and 10 unscored/beta questions), which must be completed within 60 minutes. A minimum score of 700 points on a scale of 100-900 points is required to pass the certification exam. This certification exam is a closed-book examination, so candidates are not allowed to use any notes or study materials during their examination.

To sit for the exam, you must pay an exam fee at the time of booking by purchasing an exam voucher from AKYLADE's website (https://www.akylade.com), or by purchasing a discounted exam voucher through one of our Authorized Training Partners (ATPs) as part of a bundled training package. Please visit https://www.akylade.com/partners to view a current list of Authorized Training Partners.

The certification exam can be taken online through AKYLADE's testing partner, Certiverse (https://www.certiverse.com), through their online web proctoring service from the comfort of your home or office.

AKYLADE CERTIFIED CYBER RESILIENCE PRACTITIONER (A/CCRP)

The AKYLADE Certified Cyber Resilience Practitioner (A/CCRP) certification exam is an advanced-level certification for cybersecurity and information technology professionals interested in mastering cyber resiliency by implementing the NIST Cybersecurity Framework to an organization's specific situations and needs. This advanced-level certification thoroughly covers the NIST Cybersecurity Framework and how internal and external cybersecurity consultants apply it to a real-world organization using the Cyber Risk Management Action Plan (CR-MAP) process.

To pass the AKYLADE Certified Cyber Resilience Practitioner certification exam, you must demonstrate that you can implement and apply all the material aspects of the NIST Cybersecurity Framework, including:

- How to coordinate with management for organizational buy-in and how to establish risk profiles for organizations

- How to discover top organizational cybersecurity risks using rigorous prioritization methods

- How to create a personalized cybersecurity risk management strategy tailored to an organization's unique requirements

- How to conduct maintenance and updates to the organization's cybersecurity risk posture and how to perform continuous improvement

The AKYLADE Certified Cyber Resilience Practitioner (A/CCRP) certification exam consists of 40 multiple-choice questions (30 graded questions and 10 unscored/beta questions), which must be completed within 120 minutes. These questions are all based on case studies from real-world organizations. You will be asked to analyze their organization and make recommendations to improve their cyber resiliency based on your knowledge of the NIST Cybersecurity Framework and the Cyber Risk Management Action Plan (CR-MAP) process. A minimum score of 700 points on a scale of 100-900 points is required to pass the certification exam. This certification exam is a closed-book examination, so candidates are not allowed to use any notes or study materials during their examination.

To sit for the exam, you must pay an exam fee at the time of booking by purchasing an exam voucher from AKYLADE's website (https://www.akylade.com), or by purchasing a discounted exam voucher through one of our Authorized Training Partners (ATPs) as part of a bundled training package. Please visit https://www.akylade.com/partners to view a current list of Authorized Training Partners.

The certification exam can be taken online through AKYLADE's testing partner, Certiverse (https://www.certiverse.com), through their online web proctoring service from the comfort of your home or office.

EXAM TIPS AND TRICKS

Before we dig into the contents of the AKYLADE Certified Cyber Resilience Fundamentals (A/CCRF) and the AKYLADE Certified Cyber Resilience Practitioner (A/CCRP) certification exams, it is important for you to understand some exam tips and tricks that will help you improve your performance on these exams. By understanding these tips and tricks, you can better grasp how to study for the exam as you read through the rest of this textbook. It will help you focus your efforts to get the most out of this material.

The most important thing to remember when taking the certification exams is that there are no trick questions on test day. Every question is precisely worded to match the material that you are about to study in this textbook.

During the exam, you should read each question multiple times to ensure that you understand exactly what it's asking and that you are answering the question being asked. Anytime you see the words ALWAYS or NEVER in an answer, think twice about selecting it. As in most things in life, rarely is there a case where something ALWAYS or NEVER applies to a given situation when using the NIST Cybersecurity Framework and the Cyber Risk Management Action Plan (CR-MAP) process.

As you read the questions and answers, you should be on the lookout for distractions or red herrings. Generally, at least one of these is listed in the possible answer choices to try and distract you from the correct answer. If you can identify the distractor, you can increase your chances of guessing the correct answer from the remaining answer choices provided.

Also, if you see part of a question with bold, italics, or all uppercase letters, you should pay close attention to those words because the writers of the exam questions decided that those keywords are important to selecting the correct answer.

For the AKYLADE Certified Cyber Resilience Fundamentals certification, you can rely on your knowledge of the NIST Cybersecurity Framework version 2.0 from this textbook, a video training program, or the official NIST publication as the certification ties directly to the NIST publication, "The NIST Cybersecurity Framework (CSF) 2.0" published February 26, 2024.

When you sit for the AKYLADE Certified Cyber Resilience Practitioner you should only rely on official sources, such as this textbook or training received from one of AKYLADE's Authorized Training Partners (ATP) certified to provide training for this certification exam. The reason for this is that the AKYLADE Certified Cyber Resilience Practitioner exam relies heavily on the implementation of the NIST Cybersecurity Framework using the Cyber Risk Management Action Plan (CR-MAP) process.

The CR-MAP process is not outlined or mentioned within the NIST Cybersecurity Framework itself or any of the NIST publications. This is because the CR-MAP is a proprietary framework and practical methodology created and used by Cyber

Risk Opportunities LLC to help operationalize the NIST Cybersecurity Framework and apply it to the daily operations of an organization during your future work as a cybersecurity practitioner or consultant.

Therefore, it is important to remember what concepts in the NIST Cybersecurity Framework and the Cyber Risk Management Action Plan process were covered in this textbook or an officially approved training course through an Authorized Training Partner since you will see questions about this process and its application during your exam.

Remember, this textbook covers all the testable concepts within its pages, as these are the building blocks of the AKYLADE Certified Cyber Resilience curriculum and its associated certification exams. If you study the textbook properly, you will be setting yourself up for success.

On test day, you should answer the exam questions based on your knowledge of the NIST Cybersecurity Framework version 2.0 and CR-MAP process covered in this textbook.

While it is important to have practical experience, your workplace may implement the NIST Cybersecurity Framework and its concepts differently due to their own unique situation or use cases. When in doubt, you should always select the answer that most closely matches the NIST Cybersecurity Framework as discussed in this textbook. The certification examination developers relied on the official NIST Cybersecurity Framework version 2.0, this textbook, and their decades of professional cybersecurity related experience as the basis for creating the AKYLADE Certified Cyber Resilience Fundamentals and the AKYLADE Certified Cyber Resilience Practitioner certification exams.

On exam day, you should seek to select the *best* answer from the options provided. We know that sounds a bit silly, but sometimes a question may have several right answers, but one is always the *best*, or most correct, answer.

In the world of cyber resilience and cybersecurity, there is rarely a recommendation or solution that is right 100% of the time. Instead, things tend to be more situational in the real world. The AKYLADE Certified Cyber Resilience Practitioner exam will simulate placing you into a real-world situation as a cyber resiliency consultant to provide advice, recommendations, or solutions to a fictional client during your certification exam. When in doubt, choose the correct answer in the *greatest* number of situations, and you should get the question correct on the exam.

On test day, you don't have to memorize the terms of the official NIST Cybersecurity Framework publication from this textbook word for word. Instead, you must recognize the right terms from the multiple-choice options provided.

During certification exams, you will choose your answer from a multiple-choice style question instead of a fill-in-the-blank or essay question. This is an essential difference between certification testing and the tests you may have taken in high school or college. In the certification world, you just need to be able to recognize, not regurgitate, the information being asked on the exam.

As you read this textbook and study for your upcoming exam, remember that it is important to recall the keywords and definitions for the AKYLADE Certified Cyber Resilience Fundamentals exam. For the Fundamentals level exam, you will be asked to define, recall, and explain various terms and parts of the NIST Cybersecurity Framework.

But, as you move into your studies for the AKYLADE Certified Cyber Resilience Practitioner exam, you'll be focusing on the implementation and application of the NIST Cybersecurity Framework and the Cyber Risk Management Action Plan (CR-MAP) process in a variety of different situations based on real-world events and case studies provided to you. This makes the practitioner-level exam much more difficult than the fundamental-level exam since it requires a deeper understanding than simply memorizing terms or concepts and regurgitating them on test day.

SUMMARY

This book is a comprehensive guide designed to prepare readers to successfully complete the AKYLADE Certified Cyber Resilience Fundamentals (A/CCRF) and Practitioner (A/CCRP) certification exams. By focusing on the NIST Cybersecurity Framework, the text will equip an individual with no prior experience with the necessary knowledge and skills to implement, manage, and optimize the framework within an organization.

PART ONE

CYBER RESILIENCE

CHAPTER TWO

CYBERSECURITY FUNDAMENTALS

In this chapter, we are going to introduce some key cybersecurity concepts to ensure you have the necessary knowledge needed to navigate the complexities you will face while working in the cybersecurity industry.

Your understanding of the key cybersecurity terms is crucial as you begin to implement the NIST Cybersecurity Framework out in the field. Your familiarity with these terms will ensure that you can effectively communicate and collaborate with a variety of people who are involved in implementing the NIST Cybersecurity Framework by ensuring that everyone is speaking the same language so that misunderstandings and misinterpretations are reduced.

This chapter will be a review for those of you who have already passed any of the following industry certifications: CompTIA Security+, CompTIA CySA+, CompTIA PenTest+, CASP+, ISACA's Certified Information Security Manager (CISM), ISC2's Systems Security Certified Practitioner (SSCP), ISC2's Certified Information Systems Security Professional (CISSP), or any equivalencies for these cybersecurity certification exams.

WHAT IS CYBERSECURITY?

Over the years, many terms have been used to describe the process of protecting networks and the data they contain. As you enter the industry, you may hear four different terms, which sound the same, but represent a slightly different approach to protecting your organization's systems. These four terms are information security, information systems security, information assurance, and cybersecurity.

Information security refers to the protection of information and data assets from unauthorized access, use, disclosure, alteration, or destruction. It involves implementing security measures, policies, procedures, and controls to ensure information confidentiality, integrity, and availability. Information security focuses on protecting all forms of information, regardless of the technology or system used to store or transmit it.

For example, encrypting sensitive data stored on a server and implementing access controls to limit unauthorized access to a safe containing a top-secret printed report are both examples of information security measures.

Information systems security, on the other hand, specifically focuses on protecting computer systems and the associated infrastructure that store, process, transmit, and manage information. It encompasses the security measures, policies, and controls implemented to safeguard computer hardware, software, networks, and databases from unauthorized access, attacks, and disruptions. Information systems security aims to ensure the availability, integrity, and confidentiality of information processed by computer systems.

An example of information systems security is the deployment of firewalls, intrusion detection systems, and antivirus software to protect a company's network infrastructure and systems from external threats.

Information assurance, also known as IA, is a new term in the computer security field that arose over time. Information assurance is a broader concept encompassing the management and protection of information assets, including information security and information systems security. It emphasizes the holistic approach of ensuring confidentiality, integrity, availability, and non-repudiation of information. Information assurance goes beyond technical controls and includes integrating people, processes, and technology to address risks related to information.

Information assurance also involves implementing policies, procedures, training, and risk management frameworks to ensure the proper handling and protection of information throughout its lifecycle. An example of information assurance is the development of a comprehensive information security program that includes security policies, regular risk assessments, security awareness training, incident response planning, and ongoing monitoring.

Cybersecurity is a term that has gained significant prominence in recent years and is often used interchangeably with information security. Cybersecurity specifically focuses on protecting computer systems, networks, and digital information from cyber threats, which include unauthorized access, cyber attacks, data breaches, and other malicious activities conducted through digital means. Cybersecurity involves a combination of technical, operational, and managerial measures to govern, identify, protect, detect, respond to, and recover from cyber incidents.

Cybersecurity is preferred over information assurance when addressing the unique challenges posed by the interconnectedness and digital nature of modern technology. Some examples of cybersecurity measures include implementing multi-factor authentication, conducting regular vulnerability assessments, and establishing incident response plans.

Many of these terms sound quite similar, but there are some distinctions between them that you should be aware of. Information security and information systems security have several overlapping areas of focus, with information security encompassing a broader scope that includes both information and the systems that process it. Information systems security, on the other hand, specifically focuses on protecting just computer systems and associated infrastructure.

Information assurance, though, is considered a more comprehensive approach that incorporates both information security and information systems security, emphasizing the integration of people, processes, and technology to manage an organization's information risks. Finally, cybersecurity is the most recently used term, and it emphasizes protecting against cyber threats that are specific to the digital realm to address the unique challenges posed by technology interconnectivity and digital attacks.

It is important to understand the distinction between these three terms from a theoretical perspective because many cybersecurity certification exams will focus on these distinctions in their questions. In the real world, you will often see these terms used interchangeably by practitioners in the field, or one term being preferred over another based on the practitioner's previous work experience.

The changing of these terms over time has also affected higher education's naming schemas for their degrees. For example, from 2008-2015, most degrees in computer security were termed information assurance, but since 2015 most degrees now opt to use the term cybersecurity. Similarly, any degrees in this area of study earned before 2008 were almost exclusively termed as information systems security or the even older term, computer security.

THE CIANA PENTAGON

The **CIANA pentagon** refers to the five core principles of cybersecurity that form the foundation for protecting digital assets and maintaining secure environments: confidentiality, integrity, availability, non-repudiation, and authentication. As an aspiring cybersecurity consultant, it is imperative that you understand these core principles as you apply the NIST Cybersecurity Framework within an organization.

Confidentiality in cybersecurity refers to the protection of sensitive information from unauthorized access or disclosure by ensuring that only authorized individuals or entities can access and view confidential data.

For example, encrypting sensitive customer data stored in a database ensures confidentiality by rendering the information unreadable without the proper decryption key, even if an attacker steals it. By safeguarding confidential information, organizations can mitigate the risk of data breaches and unauthorized disclosures, thereby maintaining the trust of their stakeholders and meeting compliance requirements associated with various privacy regulations.

Integrity ensures that data remains accurate, consistent, and unaltered throughout its lifecycle by protecting it against unauthorized modification, deletion, or corruption.

For instance, cryptographic hashing algorithms can be used to detect changes in data by generating unique hash values known as a hash digest. By comparing the hash digest of the original data with the recalculated hash digest, integrity violations can be detected, ensuring the data's trustworthiness, and preventing tampering or unauthorized modifications to the critical data contained within your information systems.

Availability refers to the accessibility and usability of digital assets and services when needed. In cybersecurity, ensuring availability means protecting against disruptions or denial of service that may render systems or resources inaccessible to legitimate users.

For example, implementing redundant systems and robust backup strategies can mitigate the impact of hardware failures, natural disasters, or cyber attacks. This can ensure that critical services and resources remain available and operational for your organization's authorized users.

Non-repudiation is the assurance that the originator of a digital communication or transaction can neither deny their involvement nor the authenticity of the data being exchanged.

In cybersecurity, cryptographic techniques such as digital signatures can provide non-repudiation by using a hashing algorithm to generate a unique hash value for the message and then encrypting that value with the sender's private key. By digitally signing a document or message using this digital signature, the sender can prove their identity by encrypting the hash digest with their private key and ensuring the integrity of the information being exchanged through the hash digest itself, thereby preventing any subsequent denial of their involvement in the data exchange.

Authentication verifies the identity of individuals or entities attempting to access digital systems or resources. It ensures that only authorized users gain entry and prevents unauthorized access by impostors or malicious actors. In cybersecurity, authentication methods include passwords, biometrics, and multi-factor authentication.

For instance, requiring users to provide a unique username and password, along with a fingerprint scan or a one-time verification code sent to their mobile device, would be considered a form of multi-factor authentication. Multi-factor authentication is considered the strongest form of authentication currently. It is used to thwart any attempted unauthorized access by an attacker.

The CIANA pentagon encapsulates five foundational pillars of cybersecurity: confidentiality, integrity, availability, non-repudiation, and authentication. In order for you to safeguard digital assets, you will need to have a comprehensive understanding of these principles. In short, remember that confidentiality protects sensitive information, integrity ensures data remains unaltered, availability ensures resources are accessible, non-repudiation prevents denial of involvement, and authentication verifies user identities. By mastering these concepts over time, you will be well-equipped to protect digital systems, preserve privacy, and ensure the security of digital assets in an ever-evolving threat landscape.

CYBERSECURITY INCIDENTS

A **cybersecurity incident** is any unauthorized or malicious event that compromises the confidentiality, integrity, or availability of an organization's digital assets, systems, or networks. In 2022, the average cost of cleaning up a data breach was $4.2 million per cybersecurity incident that occurred at companies worldwide. Each of these cybersecurity incidents results from a vulnerability being exploited by a threat.

Often, people who are new in the cybersecurity industry will use the words threat and vulnerability interchangeably. However, they are not technically the same thing; you must know the difference between them.

A **threat** is defined as any potential source or actor that has the capability to exploit a vulnerability, weakness, or flaw that causes harm to an organization's digital systems, networks, or data. An even more generic way to think about this is that a threat is a person or event that has the potential to impact a valuable resource in some kind of negative manner. So, cybercriminals or nation-state actors might be a threat if they wish to steal your organization's confidential data, but a hurricane is also a threat because it could cause a power outage that would render your network and systems unusable.

A **vulnerability**, on the other hand, refers to a weakness or flaw in a system, network, or software that a threat actor can exploit to compromise the security and integrity of digital assets. What makes something a vulnerability is a quality or characteristic within a given resource or its environment that might allow the threat to be realized.

If there is any weakness in the system design, implementation, software code, or lack of preventative mechanisms within an organization's information systems, then a vulnerability exists within them. For example, if the organization utilizes an end-of-life version of Microsoft Windows on its file servers, this would be classified as a vulnerability.

Similarly, if the organization only has a battery backup system for their servers that would last only 15 minutes during a power outage, then the organization is vulnerable to power outages that could be caused by inclement weather. This vulnerability can be mitigated by implementing a longer-term power generation capability, like a diesel generator, to power the backup power. But, if no longer-term power generation capability exists within the organization, then a hurricane could cause the system to lose power completely after only 15 minutes.

RISK

Threats and vulnerabilities are directly linked to determining the amount of risk that an organization faces. **Risk** refers to the potential for loss, damage, or harm resulting from the occurrence of threats exploiting vulnerabilities in digital systems or assets.

In fact, risk is used to measure the likelihood and impact of a given threat exploiting a given vulnerability. This is expressed mathematically as a formula where risk equals the threat multiplied by the vulnerability.

$$Risk = Threat \times Vulnerability$$

If the threat increases while the vulnerability remains the same, then the overall risk will increase.

$$Risk (\uparrow) = Threat (\uparrow) \times Vulnerability (\leftrightarrow)$$

If the vulnerability increases while the threat remains the same, then the overall risk will increase.

$$Risk (\uparrow) = Threat (\leftrightarrow) \times Vulnerability (\uparrow)$$

If an organization wants to keep its risk at a given level, then as the vulnerability increases, countermeasures must be put into place to reduce the threat of exploitation.

$$Risk (\leftrightarrow) = Threat (\downarrow) \times Vulnerability (\uparrow)$$

On the other hand, if the threat increases, then the organization must reduce the vulnerability's exposure to maintain the same level of risk.

$$Risk (\leftrightarrow) = Threat (\uparrow) \times Vulnerability (\downarrow)$$

In order to have risk, you must have both a threat and a vulnerability. This is obvious when you look at the mathematical formulas presented because if either the threat or vulnerability is zero, the risk would also equal zero.

Let's consider an example to put this concept into perspective. Assume that you are an iPhone user, and you just heard about a new piece of malware that is infecting smartphones. This threat concerns you, so you do some additional research and discover that this piece of malware (threat) can only be used against the Android operating system (vulnerability). Since you are running iOS on an iPhone, your system has no vulnerability to this threat. Therefore, your risk is zero regarding this specific piece of malware, so you do not need to worry about it.

Conversely, let's pretend you are the only person living on Mars. If you are worried that someone might steal your laptop because you left the front door to your Martian home unlocked, fear not because this vulnerability cannot be exploited as there is no threat. As there is no one else on the entire planet, there is no threat actor to exploit the vulnerability. Since there is zero threat, there is also zero risk.

The bottom line is that for a risk to exist, you must have both a threat that can exploit a given vulnerability, and the vulnerability itself must be present in the organization's systems. If there is no threat to exploit a vulnerability, or there is no vulnerability for the threat actor to exploit, then there cannot be a risk, negative consequence, or cybersecurity incident. This is how threats and vulnerabilities are linked together, and this is a very important concept to understand as you begin your journey into cybersecurity.

SUMMARY

In this chapter, we explored some essential cybersecurity concepts as we laid the foundation for the rest of your journey into cyber resiliency. By understanding key terms such as information security, information systems security, information assurance, and cybersecurity, you should have gained some insights into the nuances and distinctions of the language used by practitioners in the field.

The CIANA pentagon was also introduced, comprising the five core principles of confidentiality, integrity, availability, non-repudiation, and authentication. These five principles form the bedrock of cybersecurity and ensure the protection and secure management of digital assets within our organizations. Additionally, we examined the relationship between threats and vulnerabilities, emphasizing the fact that both elements must be present for risk to exist.

By mastering the key concepts covered in this chapter, you will have acquired a solid understanding of the fundamentals required to excel in the cybersecurity field. The knowledge gained on cybersecurity principles, threat and vulnerability interactions, and a basic understanding of what creates a risk to an organization will lay a strong foundation for your effective implementation of cybersecurity measures in the future. As we move forward in subsequent chapters, this understanding will serve as a valuable framework for applying the NIST Cybersecurity Framework and addressing the complex challenges of securing digital environments.

CHAPTER THREE

RISK MANAGEMENT FUNDAMENTALS

In the previous chapter, we introduced the concept of risk. Before delving into the NIST Cybersecurity Framework, it is crucial to establish a solid understanding of risk management fundamentals. This comprehensive overview will equip you with the necessary knowledge and terminology for navigating the cybersecurity industry's intricate landscape. By grasping the key concepts and principles of risk management, you will be well-prepared to effectively identify, assess, and mitigate risks, ensuring the security and resilience of digital systems and assets.

Before effectively identifying and prioritizing cybersecurity risks to support an organization's risk appetite and strategic objectives, you must first gain an understanding of proper risk analysis, risk assessment, and risk mitigation strategies. Your knowledge of risk management fundamentals will allow you to also make informed recommendations or decisions throughout the implementation process by evaluating the cost-effectiveness of different cybersecurity measures, determining appropriate risk responses, and prioritizing resource allocations.

THE RISK MANAGEMENT LIFECYCLE

Risk management is considered to be a fundamental process in achieving cyber resilience. **Risk management** is the systematic process of identifying, assessing, prioritizing, and mitigating potential risks to an organization's digital systems, networks, data, and assets to ensure their confidentiality, integrity, and availability. The conduct of proper risk management processes is crucial to effectively manage an organization's risks within the ever-changing cybersecurity landscape.

Risk management and its associated processes form what is referred to as the risk management lifecycle. The **risk management lifecycle** provides a systematic and iterative approach to managing risks by encompassing several phases: risk identification, risk assessment, risk response planning, risk mitigation, and ongoing risk monitoring and review.

The risk management lifecycle ensures that risks are continually assessed, prioritized, and addressed in a structured manner. The lifecycle enables organizations to adapt their risk management strategies based on evolving threats, business environment changes, and the effectiveness of implemented controls. Organizations can establish a proactive and resilient approach to managing risks by following the risk management lifecycle.

PHASE ONE: RISK IDENTIFICATION

The risk management lifecycle begins with risk identification, where potential risks are identified through various methods such as risk assessments, threat intelligence, and stakeholder input. The term **stakeholders** refer to an individual or group with an interest or influence in the organization's digital systems and assets, whose perspectives and requirements may shape risk management strategies and decisions.

For example, the Chief Executive Officer (CEO), Chief Operating Officer (COO), Chief Financial Officer (CFO), and Chief Information Security Officer (CISO) are all examples of key stakeholders in most organizations, but so are the organization's system administrators, business users, and other employees who rely on the organization's digital systems on a daily basis.

Another key stakeholder that coordinates with your organization may be your suppliers. A **supplier** is an external entity that provides goods, services, or resources to an organization. Assessing the associated risks with suppliers is crucial to ensure they meet the organization's security and compliance requirements, minimizing potential vulnerabilities and threats introduced through their products or services. For example, if your organization uses Amazon Web Services' (AWS) cloud-based infrastructure, then Amazon is both your supplier and a key stakeholder to consider as you begin to identify and manage your organization's risk profile.

The risk identification phase involves systematically identifying vulnerabilities, threats, and potential consequences that could impact the organization's objectives. During the risk identification phase, it is also important to work with the organization's key stakeholders to determine the organization's risk appetite and risk tolerance.

An organization's **risk appetite** refers to the organization's willingness to accept potential risks related to its digital systems and assets, guiding decision-making processes to align risk management strategies with business objectives and priorities.

For instance, an organization with a low-risk appetite, such as a financial institution handling sensitive customer data, may prioritize extensive security controls and stringent compliance measures to minimize the likelihood of data breaches. This organization might invest heavily in robust encryption mechanisms, implement strict access controls, and regularly conduct vulnerability assessments to ensure a high level of protection.

In contrast, a technology-based startup operating in the fast-paced Silicon Valley environment may have a higher risk appetite and be more willing to accept innovation and rapid growth risks. Due to this higher risk appetite, the organization might adopt agile development methodologies, embrace emerging technologies, and allocate resources to explore cutting-edge solutions while acknowledging that some vulnerabilities or risks may arise as a result of this more dynamic approach.

An organization's **risk tolerance**, on the other hand, is defined as the level of risk that an organization is willing to accept in pursuit of its objectives before action is deemed necessary to reduce it. Risk tolerance focuses on the boundaries of risk that an organization can withstand without compromising its strategic objectives. In contrast, risk appetite refers to the overall amount of risk that an organization is willing to pursue or retain to achieve its goals and this represents a broader, more strategic view of the risks the organization is prepared to take on as part of its business operations.

This is why it is important to understand the risk appetite and risk tolerance of any organization you might be working in. Your recommendations to help manage risk need to be in alignment with the organization's risk appetite so that you can align the appropriate risk management strategies to that organization to ensure a balance between increasing system security, growing their business, and their overall strategic objectives.

During the risk identification phase, each identified risk should be recorded in a risk register. A **risk register** is a centralized document or database that systematically records and tracks identified risks, along with their attributes, **assessment** results, and corresponding risk management actions, to facilitate effective risk monitoring and mitigation. This risk register is created initially during the risk identification phase, while the remainder of the information is added to each risk as the organization works through the remaining phases of the risk management lifecycle.

PHASE TWO: RISK ASSESSMENT

Once the risks have been identified, the next phase is conducting a risk assessment. During the risk assessment, risks are analyzed and evaluated to determine their likelihood of occurrence and their potential impact. By conducting a comprehensive risk analysis, organizations gain insights into the significance and prioritization of any risk that was identified in the risk identification phase.

During the risk assessment phase, the risk analysis process plays a pivotal role in understanding and quantifying the identified risks. Risk analysis involves evaluating the likelihood of a risk occurring and assessing its potential impact on the organization's digital systems and assets. There are different approaches to conducting risk analysis, including the use of qualitative, quantitative, and hybrid methods.

In **qualitative risk analysis**, risks are assessed based on subjective judgments, such as the likelihood and impact of a risk using a scale instead of numerical metrics or figures. The term **likelihood** refers to the probability of a risk event occurring, while **impact** refers to the magnitude of its consequences.

By assigning qualitative values using a scale such as low, medium, or high for each of these factors, an organization can gain a broad understanding of each risk's significance and prioritize them accordingly within the risk register.

For example, a high likelihood and high impact risk would require immediate attention and mitigation measures, while a low likelihood and low impact risk may be considered a lower priority. These resulting prioritizations can also help the organization determine which risks should receive more or fewer resources to be mitigated or resolved.

Quantitative risk analysis, on the other hand, involves evaluating risks using numerical values and metrics to assess the financial impact and frequency of risk events. This approach allows organizations to assess risks in a more objective and measurable manner. Some common values used with quantitative risk analysis are the Single Loss Expectancy (SLE), the Annualized Loss Expectancy (ALE), and Annualized Rate of Occurrence (ARO).

Single loss expectancy (SLE) is a metric used to estimate the potential financial loss an organization may experience from a single risk event occurrence. The single loss expectancy is equal to the asset value (AV) multiplied by the exposure factor (EF).

$$SLE = AV \times EF$$

The **asset value** represents the financial worth of the asset at risk, while the **exposure factor** represents the percentage of loss that would occur if the asset were compromised. For example, if an organization's web server has an asset value of $100,000 and the exposure factor for a specific risk is determined to be 60%, the SLE would be $60,000 ($100,000 x 0.60).

This means that in the event that this risk materializes, the organization could potentially face a financial loss of $60,000. Understanding the SLE allows organizations to prioritize their risk mitigation efforts based on the potential financial impact of each risk.

Annualized loss expectancy (ALE) is a metric used to estimate the expected financial loss over a specified time period resulting from a particular risk. The annualized loss expectancy is equal to the single loss expectancy multiplied by the annualized rate of occurrence.

$$ALE = SLE \times ARO$$

The **annualized rate of occurrence** (ARO) is a crucial metric in cybersecurity that represents the estimated frequency at which a specific risk event is expected to occur within a year. The annualized rate of occurrence is used in conjunction with other risk metrics, such as the single loss expectancy and the annualized loss expectancy, to assess the potential financial impact of risks. For example, if a particular risk event is expected to occur three times every ten years, the annualized rate of occurrence would be 3/10 or 0.3.

The annualized loss expectancy can be calculated using the annualized rate of occurrence and the single loss expectancy. For example, if the SLE for a specific risk is determined to be $50,000 and the ARO is estimated to be 0.2 (meaning the risk occurs 20% of the time in a year), then the ALE would be $10,000 ($50,000 × 0.2). This indicates that, on average, the organization can expect to face a financial loss of $10,000 per year due to that specific risk.

Many organization leaders prefer to use quantitative risk analysis over qualitative risk analysis because it provides a more exact value for each risk identified. By quantifying risks in monetary terms, organizations can prioritize their mitigation efforts based on potential financial impact.

Unfortunately, it is often hard to calculate the exact value for each risk because it could take significant time and resources to identify the asset value, exposure value, and annualized rate of occurrence for each identified risk in your risk register. Therefore, in some cases, a hybrid risk analysis approach may instead be employed.

A **hybrid risk analysis** combines qualitative and quantitative approaches to assess risks by incorporating subjective judgments and numerical metrics to understand the likelihood, impact, and financial implications of the identified risks. This combined approach can provide a more comprehensive understanding of risks by leveraging the strengths of both methods. This allows organizations to consider factors beyond financial impacts, such as reputation, regulatory compliance, operational disruptions, and monetary assessments. By combining these approaches, organizations can gain a nuanced perspective on the risks' potential consequences and make informed decisions about risk mitigation strategies.

By conducting risk analysis through qualitative, quantitative, or hybrid approaches, organizations can effectively evaluate and prioritize risks identified during the risk identification phase. This process enables them to allocate resources, implement appropriate risk management measures, and focus their efforts on mitigating risks that pose the most significant threats to their cybersecurity posture and overall resilience.

PHASE THREE: RISK RESPONSE PLANNING

After completing the risk assessment, the organization will begin the risk response planning phase. Organizations develop strategies and action plans during this phase to address their identified risks. Risk response actions include risk acceptance, risk avoidance, risk transference, or risk mitigation.

Risk acceptance is a risk response action that involves acknowledging the existence of a risk and choosing not to take further action to avoid, transfer, or mitigate it. Organizations may opt for risk acceptance when the cost of implementing risk mitigation measures outweighs the potential impact of the risk.

For example, a small business may accept the risk of a minor data breach due to limited resources and instead focus on investing their limited resources in their core business operations. While risk acceptance does not eliminate the risk, organizations can monitor the risk and be prepared to respond if the impact exceeds the acceptable threshold based on the organization's risk appetite.

Risk avoidance is a risk response action that aims to eliminate or minimize risks by avoiding activities or situations that pose a significant threat. This can be achieved by modifying business processes, technologies, or operational practices.

For instance, an organization may choose to avoid the risk of a third-party data breach by maintaining strict in-house data storage and processing capabilities instead of relying on external service providers. Organizations avoid potential risks by eliminating their exposure to certain threats and vulnerabilities.

Risk transference is a risk response action that involves shifting the potential impact of a risk to a third party, typically through contracts, agreements, or insurance policies. An organization may transfer risk when it lacks the expertise, resources, or desire to handle certain risks internally.

For example, an organization might transfer the risk of financial losses resulting from cyber attacks to an insurance provider by obtaining a cybersecurity insurance policy. By transferring the risk, organizations can mitigate the financial impact and share the responsibility of managing the risk with a third party. As an important aside, to ensure that the insurance policy includes the appropriate coverage necessary to transfer the risk, the organization should consult an attorney to review the insurance policy in detail prior to purchasing the insurance coverage. An insurance agent's assurance that a policy provides certain coverage will not apply should you have to make a claim; instead, the terms of the specific policy will govern.

One of the authors has been a practicing business litigation attorney for over two decades, and the number one issue she dealt with on a consistent basis was the refusal of an insurance company to reimburse claims made by an insured organization due to having improper or insufficient insurance coverage. A business cannot transfer the risk if they do not purchase appropriate and comprehensive insurance policy that specifically covers the risk anticipated and for the amount of coverage intended.

For example, when businesses were mandated to close due to COVID-19, some businesses sought to make claims for "business interruption" on their commercial insurance policies. In so doing, these businesses found out the hard way that their policies either did not cover business interruptions due to viruses or declared states of emergency limitations and exclusions, or that the coverage amount was minimal and insufficient to meet their business needs for those lengthy, unanticipated, mandatory business closures. Even though they were told by an insurance agent that the policy had business interruption coverage, many businesses found out that their insurance was insufficient to provide them proper coverage and insurance benefits.

Judges and juries across the country have made it clear that a verbal assurance as to what an insurance policy may cover does not bear weight against the coverage as defined by the actual, written terms of the insurance policy, signed by both the insured and the insurer. The written terms of the insurance contract will control, so be sure to read them and understand them. Insurance policy terms are often complex and sometimes incomprehensible. Therefore, you should make sure to consult your licensed attorney to ensure you have the right coverage in place (both in terms of scope and financial compensation) to protect your organization and to properly transfer the risk to your insurance company.

Risk mitigation is a risk response action that focuses on reducing the impact or likelihood of a risk event by implementing controls, safeguards, and countermeasures. Mitigation measures can include technical solutions, process improvements, employee training, and policy enforcement.

For instance, an organization might mitigate the risk of unauthorized access to its network by implementing multi-factor authentication, encryption, and regular security patch updates on its systems. Risk mitigation aims to minimize the potential consequences of risk and enhance the organization's resilience to cyber threats.

By considering these different risk response actions, organizations can make informed decisions on how to address risks based on their risk appetite, available resources, and the specific characteristics of each risk. It is essential to select the appropriate risk response strategy for each risk to optimize the allocation of resources and protect the organization's critical assets and operations.

PHASE FOUR: RISK MITIGATION

After completing the risk response planning phase, organizations should move into the risk mitigation phase. The risk mitigation phase occurs when the organization implements its chosen risk response strategies by deploying appropriate controls, policies, procedures, and technical measures to reduce vulnerabilities and minimize the impact of potential threats. Risk mitigation is a crucial component of the risk management lifecycle as it aims to actively reduce the likelihood and severity of risks, thereby enhancing the organization's overall cybersecurity posture and resilience.

During the risk mitigation phase, organizations will identify specific measures and actions to address the identified risks effectively. This involves a systematic and comprehensive approach to implementing controls and safeguards tailored to the unique characteristics of each risk. These measures are designed to strengthen the organization's security infrastructure, enhance incident response capabilities, and protect critical assets and systems from potential threats.

The implementation of technical controls and measures is a key aspect of risk mitigation which can involve the deployment of firewalls, intrusion detection systems, encryption technologies, access controls, and other security solutions that help prevent unauthorized access, detect anomalies, and protect sensitive data. Technical controls play a vital role in reducing vulnerabilities, fortifying network perimeters, and ensuring secure configurations of hardware and software systems.

Alongside technical controls, though, an organization must also establish and enforce robust operational procedures and policies. This includes defining clear guidelines for access management, incident response, data handling, change management, and employee awareness and training programs. By fostering a strong security culture and ensuring adherence to established procedures, organizations can mitigate risks arising from human error, negligence, or malicious intent and maintain a resilient cybersecurity environment.

PHASE FIVE: RISK MONITORING AND REVIEW

The final phase of the risk management lifecycle is ongoing risk monitoring and review. This phase involves continuously monitoring the effectiveness of risk mitigation measures, identifying new risks, and reassessing existing risks as the business and threat landscape evolve. Regular reviews and updates to the risk register ensure that risk management strategies remain aligned with the organization's objectives and risk appetite.

An organization's continual monitoring and assessment can ensure the effectiveness of its implemented controls. The organization should conduct regular vulnerability scans, penetration testing, and security audits to identify emerging risks or potential weaknesses. By proactively identifying and addressing vulnerabilities and evolving threats, organizations can adapt their risk mitigation strategies and prevent potential risks from being exploited.

By adhering to the risk management lifecycle, organizations can establish a robust and proactive approach to risk management. This enables them to enhance their overall cyber resilience by effectively identifying, assessing, and mitigating risks within their systems.

INHERENT RISK AND RESIDUAL RISK

When working within the risk management lifecycle, it is essential to understand the concepts of inherent risk and residual risk. Inherent and residual risks allow an organization to make more informed decisions and allocate resources effectively throughout the risk management lifecycle processes. This enables them to focus on reducing risks to an acceptable level and ensure their systems and operations' ongoing security and resilience.

Inherent risk refers to the level of risk in an organization's systems or processes without any control measures or risk mitigation efforts. This type of risk represents the potential impact and likelihood of a risk event occurring before any risk mitigation actions have been taken.

For example, a company's inherent risk of a data breach might be high if they store sensitive customer information without encryption or proper access controls in place.

Residual risk, on the other hand, refers to the level of risk that remains after implementing risk mitigation measures. This type of risk represents the risk that persists despite the organization's efforts to reduce it through controls and safeguards. Residual risk considers the effectiveness of the implemented risk response strategies in reducing the likelihood and impact of risks.

For instance, even after implementing encryption and access controls, a company may still have a residual risk of a data breach due to the possibility of an insider threat, an advanced persistent threat, or an emerging vulnerability that was just discovered.

Differentiating between inherent and residual risks is crucial in the risk mitigation process. By assessing inherent risk, organizations understand the baseline risk landscape. They can identify areas where significant vulnerabilities or threats exist. This information helps inform the selection and implementation of appropriate risk response strategies.

After implementing these strategies, organizations evaluate the residual risk to determine if it falls within acceptable levels based on their organization's risk appetite and strategic objectives. If the residual risk is deemed too high, additional risk mitigation measures may be necessary to further reduce the risk to an acceptable level.

BUSINESS IMPACT ANALYSIS

A **business impact analysis** (BIA) is a critical process in risk management that examines the potential impacts of disruptions on an organization's systems, processes, and operations. It involves a systematic evaluation to identify and prioritize critical systems and functions, assess their dependencies and interdependencies, and establish recovery objectives. By conducting a comprehensive business impact analysis, an organization will gain valuable insights into the potential consequences of disruptions and develop strategies to minimize their impact, enhance resilience, and ensure the continuity of essential business activities.

There are several important terms associated with a business impact analysis that cybersecurity professionals should be aware of, including recovery time objective, recovery point objective, mean time to recover, mean time between failures, single point of failure, mission essential functions, and critical systems.

The **recovery time objective** (RTO) is the targeted duration within which a business process or system must be restored after a disruption to avoid significant impacts. It defines the maximum tolerable downtime for a specific process or system.

For example, an e-commerce website may have an RTO of four hours, meaning that it must be back online within four hours of an incident to minimize financial losses and customer dissatisfaction.

The **recovery point objective** (RPO) determines the maximum acceptable amount of data loss that an organization can tolerate. It identifies the point in time to which data must be recovered following a disruption.

For instance, a financial institution may have an RPO of one hour, meaning that the recovery process should restore data up to the latest available backup taken within the last hour to ensure minimal data loss.

The **mean time to recover** (MTTR) represents the average time required to restore a failed system or process to full functionality after an incident. This metric measures the efficiency of the organization's recovery process. Organizations strive to minimize MTTR to reduce the duration of service disruptions.

For instance, if a critical system experiences a failure, the system administration team may work to ensure the MTTR is less than two hours to minimize the impact on business operations across the organization.

The **mean time between failures** (MTBF) is the average duration between two consecutive system or component failures. This metric is used to quantify the reliability and availability of a system. A longer MTBF indicates a more reliable system. In comparison, a shorter MTBF suggests a higher frequency of failures and potential disruptions.

A **single point of failure** (SPOF) refers to a component or resource that, if it fails, would cause a complete failure of an entire system or process. It represents a vulnerability that can significantly impact operations.

For example, if a critical server is the single point of failure for an organization's network, its failure would result in a complete network outage.

Mission essential functions (MEFs) are the key activities or processes that an organization must perform to maintain its core operations and fulfill its mission. Identifying MEFs is crucial in prioritizing resources and developing recovery strategies.

For example, a financial institution's mission's essential functions might include processing customer transactions, maintaining account balances, and ensuring regulatory compliance.

A **critical system** is one whose failure or disruption would significantly impact the organization's ability to deliver essential services or fulfill its mission. Identifying critical systems involves identifying the vital systems and components for the organization's operations. Organizations can focus their risk mitigation efforts by identifying critical systems and allocating resources accordingly. If a system is deemed critical, it naturally should receive more resources and attention than one that is not considered critical.

By conducting a thorough business impact analysis, an organization can better understand the potential consequences of disruptions, establish its recovery objectives, and prioritize its risk mitigation efforts. By defining the RTO, RPO, MTTR, and MTBF, addressing single points of failure, identifying mission essential functions, and recognizing critical systems, organizations can enhance their preparedness and resilience in the face of disruptions, ensuring the continuity of their operations while minimizing the impact to their stakeholders.

FINANCIAL ANALYSIS

Financial analysis is a crucial aspect of risk management that focuses on assessing the financial implications and considerations associated with cybersecurity measures and investments. It involves evaluating the costs, returns, and financial performance of cybersecurity initiatives within an organization. Three key financial metrics used in financial analysis are total cost of ownership, return on assets, and return on investment.

The **total cost of ownership (TCO)** represents the overall cost associated with owning, operating, and maintaining a particular asset or investment over its entire lifecycle. This total cost of ownership encompasses the direct and indirect costs associated with implementing and managing cybersecurity measures, such as acquiring security technologies, training personnel, monitoring systems, and responding to incidents.

The total cost of ownership should consider both upfront costs and ongoing expenses, including hardware, software, personnel, training, and maintenance. Organizations can make informed decisions regarding the cost-effectiveness of different cybersecurity investments and solutions by understanding the potential total cost of ownership.

The **return on assets (ROA)** metric is a financial ratio that measures the efficiency and profitability of an organization's use of its assets to generate earnings. Return on assets is calculated by dividing the organization's net income by its average total assets. This metric can be used to evaluate the effectiveness of cybersecurity investments in protecting and preserving the value of an organization's assets. A higher return on assets indicates a more efficient use of assets to generate returns, while a lower return on assets suggests potential inefficiencies or inadequate cybersecurity measures.

Return on investment (ROI) is a financial metric that assesses an investment's profitability and financial benefits relative to its cost. The return on investment is calculated by dividing the net profit or gain generated by the investment by the initial investment cost and expressing it as a percentage. The ROI metric helps organizations evaluate their cybersecurity initiatives' financial impact and benefits. It helps to quantify the potential return or savings resulting from reduced losses due to security incidents, improved operational efficiencies, enhanced customer trust, regulatory compliance, and other positive outcomes. A higher return on investment indicates a more financially rewarding investment, while a lower return on investment suggests a need for further evaluation or adjustment of the cybersecurity strategy.

For example, let's consider a hypothetical scenario where a company invests in a new cybersecurity solution to mitigate the risk of data breaches. The total cost of ownership analysis would include the upfront costs of purchasing the solution, training staff, and ongoing expenses such as maintenance and updates. The return on assets analysis would evaluate how effectively the cybersecurity investment protects the organization's assets and contributes to overall profitability. Finally, the return on investment analysis would determine the financial benefits of the investment, such as reduced losses from data breaches or potential cost savings from improved operational efficiency.

Organizations can make data-driven decisions regarding their cybersecurity investments by conducting financial analysis. As a cybersecurity professional, you may be asked to justify your risk mitigation recommendations to ensure they are financially sound. For this reason, you must understand the total cost of ownership, the return on assets, or the return on investment that your proposed solution may provide if it is approved for implementation. These metrics assist your organization in evaluating the financial viability, efficiency, and effectiveness of its cybersecurity measures while ensuring that resources are allocated appropriately for maximum value.

SUMMARY

The risk management lifecycle is used to guide an organization through a systematic approach to identify, assess, respond to, and mitigate risks. Risk identification involves engaging key stakeholders and suppliers to identify potential risks and establish the organization's risk appetite. Risk assessment employs qualitative, quantitative, or hybrid methods to analyze and evaluate risks based on their likelihood and impact. Risk response planning includes actions such as risk acceptance, avoidance, transference, and mitigation to address identified risks. The risk mitigation phase focuses on implementing controls and measures to reduce vulnerabilities and minimize the impact of potential threats. Ongoing risk monitoring and review ensures that risk management strategies remain aligned with the organization's objectives and risk landscape. It is also important to remember that there are two types of risk: inherent risk and residual risk. By understanding inherent risk and residual risk, your organization can make more informed decisions and allocate resources effectively throughout the risk management process.

Business impact analysis provides insights into the potential consequences of disruptions and aids in prioritizing mission essential functions and critical systems and identifying single points of failure. By performing a business impact analysis, an organization can establish metrics for the recovery time objective, recovery point objective, mean time to recover, and mean time between failures.

Financial analysis is another key aspect of risk management. Financial analysis evaluates the costs, returns, and financial implications of cybersecurity investments, including the total cost of ownership, return on assets, and return on investment. By applying these principles and practices, organizations can enhance their cyber resilience and protect their digital systems and assets.

CHAPTER FOUR

NIST CYBERSECURITY FRAMEWORK

The NIST Cybersecurity Framework was designed to help businesses and organizations of all sizes to better understand, manage, and reduce their cybersecurity risk and protect their information systems and the data they contain. All businesses and organizations have some level of risk to their operations due to their increased reliance on information technology, operational technology, and the networks that connect them together.

To help provide a common language and systematic methodology for managing cybersecurity risk and enhancing cyber resilience, the National Institute of Standards and Technology (NIST) developed the NIST Cybersecurity Framework.

The **NIST Cybersecurity Framework** (CSF) is defined as the set of guidelines, best practices, and standards developed by the United States government based upon input by experts in the private industry to help organizations manage and improve their cybersecurity risk management processes.

DEVELOPMENT OF THE NIST CYBERSECURITY FRAMEWORK

The **National Institute of Standards and Technology (NIST)** is a government organization that exists within the United States Department of Commerce. NIST was originally created in 1901, way before computers were even imagined. NIST was established as a non-regulatory federal agency that is focused on promoting innovation and industrial competitiveness in the United States by advancing measurement science, standards, and technology in ways that enhance economic security and improve our quality of life.

On February 12, 2013, then President Barack Obama released **Executive Order 13636**, also known as Improving Critical Infrastructure Cybersecurity. This executive order aims to improve critical infrastructure cybersecurity by establishing a framework for information sharing and collaboration between the government and private sector entities.

In this executive order, the President stated, "It is the policy of the United States to enhance the security and resilience of the Nation's critical infrastructure and to maintain a cyber environment that encourages efficiency, innovation, and economic prosperity while promoting safety, security, business confidentiality, privacy, and civil liberties."

This executive order effectively established the requirements for the NIST Cybersecurity Framework and provided the initial design criteria that include the ability to:

- Identify security standards and guidelines applicable across sectors of critical infrastructure

- Provide a prioritized, flexible, repeatable, performance-based, and cost-effective approach

- Help owners and operators of critical infrastructure identify, assess, and manage cyber risk

- Enable technical innovation and account for organizational differences

- Provide guidance that is technology neutral and enables critical infrastructure sectors to benefit from a competitive market for products and services

- Include guidance for measuring the performance of implementing the CSF

- Identify areas for improvement that should be addressed through future collaboration with particular sectors and standards-developing organizations

While NIST was responsible for getting the framework created and published; the primary authors were cybersecurity practitioners from multiple organizations within a variety of industries across the United States. These practitioners met with NIST over five separate workshops held at different geographic locations across the United States to identify existing cybersecurity standards, guidelines, frameworks, and best practices that applied to increasing the security of critical infrastructure sectors and other interested entities; specify high-priority gaps for which new or revised standards were needed; and collaboratively develop action plans by which these gaps could be addressed.

The result of these workshops and the subsequent authoring process was the first version of the cybersecurity framework titled the "Framework for Improving Critical Infrastructure Cybersecurity," which was released in February 2014 as version 1.0.

The framework quickly began to be adopted inside the critical infrastructure sector, which was the original intent, and across a wide variety of industries and sectors. This led to the framework becoming more broadly accepted and known more commonly under the name of the NIST Cybersecurity Framework, the NIST CSF, or simply as CSF. These days, you can find the NIST Cybersecurity Framework being used by a large variety of companies, organizations, non-profits, and governments worldwide.

➤ On April 16, 2018, the next version of the NIST Cybersecurity Framework, version 1.1, was released. This version of the framework was designed to be backward compatible with the original framework, version 1.0. Version 1.1 was also designed with some helpful additions, including a new self-assessment section; a greater focus on supply chain risk management; and refinements were made to account for better authentication, authorization, and identity proofing outcomes.

On February 26, 2024, the latest version of the NIST Cybersecurity Framework, version 2.0, was released. This version of the framework was created with a focus on refining, clarifying, and enhancing the existing version 1.1 for better understanding between the various stakeholders who may be using the CSF in a given business or organization.

Version 2.0 also included a name change from "Framework for Improving Critical Infrastructure Cybersecurity" to the more widely accepted and commonly used nomenclature of "Cybersecurity Framework" with an official abbreviation as the CSF.

Timeline showing the more inclusive development process of CSF version 2.0 (2022-2024)

Other relevant improvements included a change in the scope of the framework. When version 2.0 was released, it ensured that the benefits of the framework can be applied to all organizations regardless of their associated sector, type, or size.

The NIST Cybersecurity Framework version 2.0 also highlights the importance of governance and supply chains to organizations' cyber resiliency. In fact, this change moved governance into its own function as part of the framework core as you will learn later on when we discuss the six functions in the NIST Cybersecurity Framework.

The final major consideration that version 2.0 incorporated is the addition of more international collaboration and engagement. Since the first release of the framework with version 1.0, it has been identified by numerous organizations that the international use of the cybersecurity framework would improve the efficiency and effectiveness of their own organization's cybersecurity efforts. While nothing is preventing a global audience from using the current version of the framework, international industry experts and foreign government representatives were not explicitly sought out in the development of versions 1.0 and 1.1. With version 2.0's development, NIST prioritized working with organizations to not just develop the cybersecurity framework's latest version, but also to develop translations into multiple languages outside of the current English editions allowing for more world-wide application.

RELEVANT EXECUTIVE ORDERS AND REGULATIONS

The NIST Cybersecurity Framework has emerged as a foundational tool in the field of cybersecurity, aiding organizations in managing and mitigating cyber risks. To understand its significance, it is crucial to explore its history and how it relates to several key executive orders issued by the President of the United States and the regulations passed by the United States Congress.

The roots of the NIST Cybersecurity Framework can be traced back to Executive Order 13636, signed by then President Barack Obama in February 2013. This executive order recognized the growing threats to critical infrastructure and called for the development of a framework to enhance the cybersecurity posture of the United States. It aimed to foster collaboration between the government and private sector entities to improve the protection and resilience of critical infrastructure.

In response to Executive Order 13636, NIST embarked on an extensive collaboration effort, engaging stakeholders from various sectors, including industry, government, and academia. This collaborative process aimed to develop a flexible and voluntary framework that could be widely adopted to help organizations manage cybersecurity risks and strengthen their resilience.

The **Cybersecurity Enhancement Act of 2014** was passed by Congress and signed into law in December 2014. This regulation aimed to strengthen and advance cybersecurity research and development efforts in the United States. The NIST Cybersecurity Framework played a significant role in this act, providing a foundational framework for organizations to align their cybersecurity efforts and adopt best practices.

The **Federal Information Security Modernization Act (FISMA) of 2014** is another important piece of legislation that contributed to the framework's relevance and use within the federal government. FISMA updated and modernized the approach to federal information security management to update and amend the older Federal Information Security Management Act of 2002. The newer FISMA emphasized the adoption of risk-based approaches and the use of industry standards, including the NIST Cybersecurity Framework, to enhance the security posture of federal agencies and improve the protection of federal information systems.

In addition to these regulations, Congress also passed the **Cybersecurity Information Sharing Act (CISA) of 2015** to facilitate the sharing of cybersecurity threat information between the government and the private sector. The NIST Cybersecurity Framework played a complementary role in this act, serving as a guide for organizations to enhance their cybersecurity practices and establish effective information-sharing mechanisms.

In May 2017, then President Donald Trump signed **Executive Order 13800**, which further reinforced the importance of the NIST Cybersecurity Framework. This order emphasized the need for executive branch agencies to implement the framework and encouraged the private sector to adopt it. Executive Order 13800 recognized the framework's value in improving risk management and prioritizing cybersecurity investments across various sectors to aid in improving the United States' overall national defense posture.

In May 2021, President Joseph Biden signed **Executive Order 14028**, which aimed to strengthen the cybersecurity of federal networks and improve information sharing between the U.S. government and the private sector on cyber threats, incidents, and risks. The goal of this executive order was to enhance the nation's ability to prevent, detect, and respond to cyber incidents. The executive order mandated that the federal government adopt best practices and operate more securely and efficiently by modernizing federal cybersecurity, enhancing software supply chain security, establishing a cybersecurity safety review board, and creating a standard playbook for responding to cyber incidents. The order also highlighted the need for improving detection of cybersecurity incidents on federal government networks and improving investigative and remediation capabilities. While this executive order did not directly mention the NIST Cybersecurity Framework by name, the executive order's practices and standards aligns closely with the principles and guidelines of the framework by reinforcing its role as a critical component of the nation's cybersecurity strategy.

Overall, the NIST Cybersecurity Framework has evolved as a pivotal tool for cyber resiliency, having been shaped by various executive orders and regulations. The framework provides us with a flexible and adaptable approach to managing cybersecurity risks, promoting collaboration, and improving the overall resilience of organizations. The NIST Cybersecurity Framework's significance is further underscored by its integration within various cybersecurity-related policies, acts, and government initiatives, reflecting its status as a widely recognized and respected framework in the field of cybersecurity.

APPLICABILITY OF THE CYBERSECURITY FRAMEWORK

The NIST Cybersecurity Framework was created by the United States government along with input from experts in the private industry, and it is now published so that anyone can use it within their own organizations free of charge. The NIST Cybersecurity Framework is provided as a public service and is considered to be public information that may be fully distributed or copied for your own organization's use without paying any licensing fees. This free tool is extremely valuable because it can be quickly implemented to provide your organization with an instant return on its investment at little to no upfront cost.

Over the past decade, doing business on the internet has become an essential part of our global economy and a huge growth driver for organizations and businesses alike. The internet has allowed small businesses to quickly scale to serve customers around the globe by leveraging its global reach.

For example, Jason Dion, one of the authors of this textbook, established his small cybersecurity training company in 2017. Within six years, this company has successfully served over 1 million students across more than 195 countries worldwide, all from the comfort of their offices in the United States.

Despite being a small business with fewer than 30 team members, Dion Training leveraged the power of asynchronous certification training courses delivered online. This delivery mechanism enabled them to reach and accommodate hundreds of thousands of students annually, with a geographically dispersed staff spanning seven countries and ten different time zones.

By using the internet, Dion Training was and continues to be able to extend its reach far beyond what would have been possible in a traditional classroom setting with limited student capacity and associated location constraints. The impact of Dion Training's online presence has allowed it to fulfill its mission and serve a significantly larger audience than it had ever envisioned.

These days, most businesses and organizations have at least some connectivity to the internet as part of their business operations. This makes it difficult to even think back 30 years ago to a time before the internet had become so ingrained into our collective business processes and daily lives.

Unfortunately, with this increased scale and connectivity, our organizations are also more at risk than ever due to the rapid rise of cyber criminals and nation-state actors. Every year, more and larger data breaches are occurring than in the previous years, and larger and higher bandwidth distributed denial of service attacks are being attempted by these threat actors.

This exploitation of the cyber risks involved with operating your organization's business online is causing trillions of dollars in damages annually. This figure seems to be continuing to grow year after year, as well.

For example, if we look back a few years to 2015, the global cost of cyber failures and attacks of all kinds was estimated to only cost approximately $500 billion per year globally. If we fast forward just six years to 2021, the cost rose to an estimated $6 trillion per year globally.

According to experts at the time of publication of this textbook, it is estimated that this total will reach $10 trillion in damages during the 2024 calendar year. That is a 66.7% increase in just three years. That is a staggering amount. If cyber risk was considered a global economy, it would become the third largest global economy in the world after the United States and China.

These days, it isn't a matter of *if* you will become the victim of a cyber attack; it is really a matter of *when* it will occur and *how bad* that data breach and its resulting cost will become. Unfortunately, running a business in the modern economy comes with risk, and a large portion of that risk is related to cyber attacks, data breaches, and your organization's own technical implementation challenges.

So, our organizations need to be prepared against these cyber risks. Using the NIST Cybersecurity Framework, you can organize your defensive and incident response capabilities to be more resilient against cyber attacks and recover more quickly in case your organization eventually becomes victimized. This is at the heart of what you are asked to do as a cyber resiliency consultant.

CHARACTERISTICS OF THE FRAMEWORK

The NIST Cybersecurity Framework possesses several distinct characteristics that set it apart from other frameworks available in the cybersecurity industry today. An organization must understand these characteristics in order to adopt and implement the framework effectively. These characteristics include: the voluntary set of guidelines; its flexibility and adaptivity; a focus on risk instead of technical controls; a focus on risk instead of compliance requirements; its ability to facilitate communication and collaboration; and its continually improving and evolving nature.

First, it is important to note that the NIST Cybersecurity Framework is a voluntary framework, which means that its adoption and implementation are not mandatory for organizations. Instead, it provides a flexible and customizable approach that organizations can voluntarily adopt to enhance their cybersecurity posture. This voluntary nature allows organizations to tailor the framework to their specific needs, considering their unique risks, capabilities, and business objectives.

Second, the framework emphasizes flexibility and adaptivity to accommodate the diverse cybersecurity requirements of different organizations. It provides a risk-based approach, allowing organizations to assess and prioritize their cybersecurity risks based on their specific context. This flexibility enables organizations to align their cybersecurity efforts with their business goals and adapt to the evolving threat landscape and technological advancements.

Third, the framework emphasizes a focus on risk instead of technical controls. Unlike some other frameworks that primarily focus on technical controls and specific security measures, the NIST Cybersecurity Framework places a greater emphasis on risk management. It encourages organizations to identify, assess, and prioritize their cybersecurity risks, enabling them to make informed decisions about allocating resources and implementing appropriate controls.

Fourth, the framework focuses on risk and prioritizing risk management rather than compliance requirements. While compliance with regulations and standards is important, the framework encourages organizations to go beyond mere compliance by fostering a risk-based mindset. By focusing on risk, organizations can better understand their vulnerabilities, anticipate threats, and proactively address cybersecurity challenges.

Fifth, the NIST Cybersecurity Framework helps facilitate effective communication and collaboration among organizational stakeholders. It provides a common language and structure for discussing cybersecurity risks, enabling different teams and departments to communicate effectively and align their efforts. This characteristic fosters a culture of collaboration, ensuring that cybersecurity considerations are integrated into various aspects of the organization's operations.

Sixth, the NIST Cybersecurity Framework is continually improving and evolving. The framework is designed to evolve and adapt to emerging threats, technologies, and best practices by undergoing regular updates and revisions based on feedback from stakeholders and the evolving cybersecurity landscape. This characteristic ensures that the framework remains relevant and effective in addressing the ever-changing nature of cyber risks, enabling organizations to stay current with the latest cybersecurity practices.

By embracing these characteristics, organizations can leverage the NIST Cybersecurity Framework to build a robust and adaptable cybersecurity program. The framework's voluntary nature, flexibility, risk focus, and emphasis on collaboration empower organizations to proactively manage cyber risks and protect their valuable assets and operations. As organizations engage with the framework, they can contribute to its ongoing improvement and align themselves with industry-leading cybersecurity best practices.

CYBER RESILIENCE

Cyber resilience refers to an organization's ability to withstand and adapt to cyber threats by implementing proactive measures, effectively responding to, and recovering from, cyber attacks or disruptions, and maintaining essential functions while minimizing damage. This encompasses a range of strategies, including robust security controls, regular vulnerability assessments, employee education on cybersecurity best practices, and the establishment of incident response plans.

By taking proactive steps, organizations can prevent or minimize the impact of cyber incidents. However, in the event of an incident, cyber resilience is crucial for ensuring business continuity and rapid recovery. A resilient organization should be able to quickly isolate, restore, and recover its systems to their comparable state prior to the incident or cyber attack.

The benefits of cyber resilience are substantial. By effectively implementing cyber resilience measures, organizations can reduce the financial cost of incidents, minimize downtime, and protect their reputation with key stakeholders. The NIST Cybersecurity Framework provides valuable guidance for achieving cyber resilience, and the public nature of the framework means organizations can derive significant upfront value without the need for additional funding or procurement.

Consider the example of an organization with robust security controls in place, regularly testing for vulnerabilities, and well-trained employees who can swiftly detect and respond to a cyber attack. By isolating and containing the incident effectively, they can limit the damage, minimize downtime, and restore normal operations efficiently. This safeguards critical functions and helps maintain trust and confidence among customers, partners, and other stakeholders.

By embracing cyber resilience, organizations can fortify their defenses, enhance their incident response capabilities, and ensure their ability to recover swiftly from cyber incidents. This proactive approach enables them to navigate the evolving cybersecurity landscape confidently and resiliently.

CRITICAL INFRASTRUCTURE

The NIST Cybersecurity Framework was initially created for organizations and businesses that operate critical infrastructure in the United States, but it has since expanded in its usage and acceptance well beyond just those working in the critical infrastructure sectors.

The term **critical infrastructure** is defined by the United States Department of Homeland Security as any physical or virtual infrastructure that is considered so vital to the United States that its incapacitation or destruction would have a debilitating effect on security, national economic security, national public health or safety, or any combination of these.

As defined by the Department of Homeland Security, there are 16 critical infrastructure sectors:

- Chemical – Organizations and companies that manufacture, store, use, and transport potentially dangerous chemicals used by other critical infrastructure sectors

- Commercial Facilities - Buildings, facilities, and spaces used for commercial purposes, including retail, entertainment, and hospitality

- Communications - Networks, systems, and assets involved in providing communication services, including broadcasting, telecommunications, and internet service providers

- Critical Manufacturing - Facilities and processes involved in the production of essential goods, such as metals, machinery, transportation equipment, and pharmaceuticals

- Dams - Structures, systems, and resources related to dam operations and water control, including hydroelectric power generation

- Defense Industrial Base - Companies and assets involved in the research, development, production, and maintenance of defense-related equipment, systems, and services

- Emergency Services - Agencies, organizations, and personnel responsible for emergency management, firefighting, medical services, and public safety

- Energy - Resources, systems, and infrastructure involved in the production, transmission, and distribution of energy, including electricity, oil, and natural gas

- Financial Services - Institutions and systems providing financial services, including banking, insurance, investment, and payment systems

- Food and Agriculture Sector - Facilities, systems, and resources related to the production, processing, and distribution of food, beverages, and agricultural products

- Government Facilities - Buildings, offices, and structures used by federal, state, local, tribal, and territorial governments for administrative and public services

- Healthcare and Public Health - Facilities, personnel, and networks involved in providing healthcare services, medical research, and public health support

- Information Technology - Systems, networks, and infrastructure used for information processing, storage, and communication, including software development and cybersecurity

- Nuclear Reactors, Materials, and Waste - Facilities, processes, and materials associated with nuclear power generation, research, and waste management

- Transportation Systems - Infrastructure, networks, and assets involved in the movement of people and goods, including aviation, maritime, rail, and road transportation

- Water and Wastewater Systems - Facilities, systems, and resources responsible for providing drinking water and managing wastewater treatment and disposal

To help oversee the protection of the various organizations and businesses in each critical infrastructure sector, one of the U.S. government departments is assigned as the Sector-Specific Agency lead for each of the 16 sectors. For example, the Department of Health and Human Services is the assigned Sector-Specific Agency for the Healthcare and Public Health Sectors. Similarly, the Environmental Protection Agency is assigned as the Sector-Specific Agency for the water and wastewater systems sector.

Due to the increasing threat of cyber attacks that could exploit cyber risks in an organizational network, the organizations responsible for critical infrastructure need to have a modern approach to identifying, assessing, and managing cybersecurity risk. This approach needs to work regardless of the organization's size, threat exposure, or cybersecurity sophistication because threat actors aren't simply going to take it easier on your company just because it is a smaller organization.

In fact, the opposite is quite often true. Attackers will target smaller organizations because they tend to be easier targets and do not have a robust cybersecurity workforce to help them defend their systems against attack. This is one of the reasons why the NIST Cybersecurity Framework was created. The framework is designed to work for organizations of all sizes and reduce the complexity of other existing frameworks that are available for use.

INTENDED AUDIENCE FOR THE NIST CYBERSECURITY FRAMEWORK

Even though the NIST Cybersecurity Framework was initially focused on cyber risk management for organizations operating critical infrastructure, it has since been expanded for use well beyond those 16 sectors. Since the framework is flexible in nature, many other organizations and industries saw the benefit of adopting it for their own use, which in turn has led to the exponential growth in the adoption of the NIST Cybersecurity Framework.

The NIST Cybersecurity Framework provides valuable guidance to organizations in various industries. For example, it can be used to help retail organizations protect their customer's data, secure the company's online transactions, and manage their supply chain vulnerabilities. The framework is also used in manufacturing to address industrial control system security and intellectual property protection and to help secure product development. It is important to note that neither retail, nor manufacturing is classified as part of the critical infrastructure of the United States. By adopting the framework, organizations across most industries and sectors can enhance their cybersecurity practices, mitigate risks, and ensure their continuity of operations.

The framework itself can be used by organizations of any size. However, with versions 1.0 and 1.1, many practitioners believed that the framework was more relevant or skewed toward larger organizations with at least 500 employees. That said, the framework has also been adopted and implemented by small and medium-sized businesses, known collectively as the SMB market, which have organizations with somewhere between 1 and 500 employees. Additionally, as of version 2.0, NIST states that the "CSF is designed to be used by organizations of all sizes and sectors, including industry, government, academia, and non-profit organizations, regardless of the maturity level of their cybersecurity programs."

When it comes to determining who is best suited to use the NIST Cybersecurity Framework, the answer comes down to who is willing to adopt and implement it within their organization. The framework is a series of best practices and guidelines and not a compliance standard that must be strictly adhered to. This means that it can be scaled up or down in size to meet the requirements of your specific organizational needs. In fact, version 2.0 of the framework is only 32 pages long, making it quite concise and relatively quick to implement within your organization.

PURPOSE OF THE NIST CSF

The purpose of the framework is to help organizations to:

(1) Describe their current cybersecurity posture

(2) Describe their target state for cybersecurity

(3) Identify and prioritize opportunities for improvement within the context of a continuous and repeatable process

(4) Assess progress toward the target state

(5) Communicate among internal and external stakeholders about an organization's cybersecurity risk

The framework is not considered a one-size-fits-all approach to managing cybersecurity risk, but instead, it allows organizations to vary how they customize the practices and processes described in the framework to best meet their own threats, vulnerabilities, risk tolerances, and organizational size and capabilities.

In addition to the core framework, NIST also publishes implementation guidance for different industries and use cases, such as federal agencies trying to implement the framework for use within their organization or a manufacturing business trying to implement the framework within their industry.

Ultimately, organizations can determine activities important to critical service delivery and prioritize investments to maximize the impact of each dollar spent as they aim to reduce and better manage their organization's cybersecurity risk.

SUMMARY

The NIST Cybersecurity Framework (CSF) is a comprehensive set of guidelines and best practices developed by the National Institute of Standards and Technology (NIST) to assist organizations in managing cybersecurity risks and safeguarding their information systems. It has gained widespread adoption across industries and sectors since its creation in response to Executive Order 13636, written by then President Barack Obama in 2013.

The framework is flexible, scalable, and applicable to organizations of all sizes, enabling them to assess their cybersecurity posture, set target states, identify areas for improvement, measure progress, and communicate risks effectively. The NIST Cybersecurity Framework is used to emphasize resilience and to help organizations prepare for, and respond to, cyber risks while facilitating quick recovery from incidents. With the increasing frequency and cost of cyber threats, the NIST Cybersecurity Framework offers a systematic approach to cybersecurity management and aligns with relevant executive orders and regulations, supporting an organization's risk-based approaches and information-sharing initiatives.

Overall, the NIST Cybersecurity Framework is highly regarded for its practicality and effectiveness in enhancing cybersecurity practices and mitigating risks. It provides organizations with valuable guidance to protect their information systems and data, ensuring the continuity of their operations. By adopting the framework, organizations can effectively manage cybersecurity risks and bolster their resilience in the face of an ever-evolving threats landscape.

CHAPTER FIVE

FRAMEWORK COMPONENTS

Imagine that you are about to embark on a cross-country road trip, eager to explore new destinations and experiences along the way. To ensure a smooth journey, you'll need a roadmap that outlines the best routes, highlights key landmarks, and offers guidance on potential challenges. Similarly, organizations require a reliable roadmap in cybersecurity to navigate the complex landscape of cyber threats and protect their digital assets. This is where the NIST Cybersecurity Framework and its components come into play.

Just as a roadmap provides structure and direction for a journey, the framework core, framework implementation tiers, and framework profile (collectively known as the framework components) help organizations establish a strong cybersecurity foundation, assess their current posture, and tailor their security practices to meet specific goals and requirements.

Throughout this book, we will delve into the intricacies of each framework component in its own chapter in order to explore how they contribute to an organization's cyber resilience and overall cybersecurity posture. Before we do, though, it is important that you gain an overall understanding of the framework components so that you can understand how they are intricately linked together.

First, we will look at the framework core that forms the heart of the NIST Cybersecurity Framework and guides organizations in governing, identifying, protecting, detecting, responding to, and recovering from cyber incidents. Then, we will examine the framework implementation tiers that enable organizations to assess and express their cybersecurity maturity level and determine the effectiveness of their security practices. Finally, we will explore the framework profile, which is a customizable tool that allows organizations to align the framework with their unique risk management objectives, industry-specific regulations, and internal priorities.

By better understanding and leveraging these framework components, organizations can better navigate the ever-evolving cybersecurity landscape, enhance their defenses, and build a resilient cybersecurity posture that effectively safeguards their critical assets and operations. So, fasten your seatbelts as we embark on a journey through the framework components, uncovering the essential elements that empower organizations to establish a robust cybersecurity foundation and adapt to the challenges of today's digital world.

THE FRAMEWORK CORE

The framework core, commonly referred to as the **core**, is a set of common cybersecurity functions, activities, and desired outcomes (in the form of functions, categories, and subcategories) that help organizations manage and mitigate cyber risks across most businesses and organizations. The core presents industry standards, guidelines, and best practices in a way that allows communication with various stakeholders across the organization, regardless of whether they operate at the executive, operational, implementation, or tactical levels. By giving everyone a common lexicon to use when describing cybersecurity activities and outcomes, it can greatly enhance communication up and down the organizational hierarchy and ensure everyone is well understood.

The core consists of six concurrent and continuous functions that all organizations perform when conducting cybersecurity activities. The term function refers to a high-level cybersecurity grouping that combines related activities and outcomes together in order to achieve specific cybersecurity objectives. The six functions are Govern, Identify, Protect, Detect, Respond, and Recover.

For each function, there are also underlying key categories and subcategories with specific outcomes. For each subcategory, these activities are then matched with example informative references from existing standards, guidelines, and best practices used throughout the industry.

The core is one of the largest parts of the framework, consisting of 22 cybersecurity categories and 106 subcategories that are commonly found across most businesses and organizations. It is important to remember that the functions should be performed concurrently and continuously to form an operational culture that addresses the dynamic nature of cybersecurity risk. These functions are not intended to form a serial or linear path that will ultimately achieve a static desired end state. Instead, just as the adversary is constantly evolving and adapting, your organization needs to continually adopt and adapt, too.

THE FRAMEWORK IMPLEMENTATION TIERS

The framework **implementation tiers**, also known as the **tiers**, are used to provide context on how an organization perceives a given cybersecurity risk and the processes or mitigations put in place to better manage that risk. An implementation tier represents the level of effectiveness in implementing cybersecurity practices within an organization, ranging from partial to adaptive. The implementation tiers help an organization understand how well it's practicing cyber security risk management activities as described by the framework.

These tiers help understand how well an organization is aware of its risks and threats, how repeatable the outcomes they produce are, and how well the organization adapts to new risks and threats. The implementation tiers are broken down and classified using a tier number from one to four, with one being the lowest tier and four being the highest tier.

Tier 1 organizations are labeled as partial organizations. Tier 1 organizations have overall ineffective risk management methods. This tier represents an initial stage of cybersecurity implementation where organizations have limited awareness and capabilities, with cybersecurity practices being ad hoc and reactive in nature. They usually have unsystematic risk management processes, unreliable risk management programs, and unresponsive risk management participation. It is hard to think of any reason an organization would want to be labeled as a Tier 1 organization, but often that is where your organization starts out. Hopefully, that isn't their desired end state or targeted tier level you are trying to achieve.

Tier 2 organizations are labeled as risk-informed organizations. Tier 2 organizations have informal risk management methods with unfinished risk management processes, underdeveloped risk management programs, and incomplete risk management participation. They aren't as bad off as the Tier 1 organizations, but there is still a lot of room for improvement. Generally, we define a Tier 2 organization as having a higher level of cybersecurity implementation within an organization by having developed some formalized policies and procedures and with greater awareness and proactive cybersecurity practices in place.

Tier 3 organizations as labeled as repeatable organizations. Tier 3 organizations have structured risk management methods with orderly risk management processes, robust risk management programs, and routinely reviewed risk management participation. Tier 3 organizations are doing relatively well in terms of risk management, and this is often a good place for an organization to be in terms of its tier level. While there is room for improvement, overall, they are doing a good job with risk management. Many organizations will decide that reaching Tier 3 is good enough based on the cost and resources that may be required to achieve the next higher tier level. Tier 3 organizations demonstrate a proactive approach to cybersecurity, focusing on continuous improvement and the ability to respond effectively to emerging threats and vulnerabilities.

Tier 4 organizations are labeled as adaptive organizations. Tier 4 represents the highest level of cybersecurity implementation within the NIST implementation tiers. These organizations have a proactive, innovative, and adaptive approach to cybersecurity and the organization's risk management methods. These organizations demonstrate the characteristics of Tier 3 and have an advanced capability to adapt and respond to evolving cybersecurity risks. Tier 4 organizations actively seek out emerging technologies, collaborate with industry partners, and participate in research and development efforts to stay ahead of cyber threats. Tier 4 organizations continuously strive for excellence and maintain a strong cybersecurity posture that enables them to effectively protect their systems, assets, and sensitive information.

It can be very difficult to reach Tier 4 in most organizations, and it takes a significant investment in your organization's programs and people to reach this level of adaptivity. Many organizations decide it simply isn't worth the cost or effort, so they may remain at Tier 3 instead.

Remember, the organization self-assigned these implementation tiers based on where they perceive themselves currently and which tier they wish to reach in the future. The organization should consider its current risk management practices, threat environment, legal and regulatory requirements, business and mission objectives, and organizational constraints when selecting its implementation tier.

Not every organization will be considered a Tier 4 or adaptive organization, but similarly, not all organizations strive to become a Tier 4 organization in the future. Instead, think of the tiers as a clear status report of where you are and where you are going, not a graded report card from A to F like you may have had back in high school.

THE FRAMEWORK PROFILE

The framework profile, or the **profile**, represents an organization's cybersecurity objectives, current state, and target state that provide a roadmap for aligning cybersecurity activities and priorities with the organization's business requirements. This profile is created from outcomes based on an organization's business needs selected from the framework core's categories and subcategories. This profile can be considered the alignment of standards, guidelines, and best practices to the framework core in a particular implementation scenario for a given organization. Your organization will need to develop its own **organizational profile** to effectively use the NIST Cybersecurity Framework to manage and mitigate its risks. The profile can then be used to identify opportunities for improving your organization's cybersecurity posture by comparing its current profile with a target profile. This will allow you to assess where your organization is today, where it wants to be in the future, as well as the gap between those two profiles.

To develop a profile, your organization should review all the categories and subcategories in the framework and then select the relevant ones, based on its business or mission drivers and risk assessment, to determine which are the most important for your specific organization. Once you identify those categories and subcategories that are important to meet your business's objectives by reducing or mitigating risk, you can then craft an appropriate profile for your organization.

The **current profile** depicts an organization's existing cybersecurity practices, including its cybersecurity activities, desired outcomes, and current risk management approaches. The **target profile**, on the other hand, represents the organization's desired state of cybersecurity practices and outcomes and outlines the specific cybersecurity improvements and goals it aims to achieve.

Once the organization creates its current profile and target profile, it can be used to conduct a gap analysis between your current state and desired future state. The results of this gap analysis will then be used to create a plan of action. This action plan should use proper prioritization based on the profiles you created and factor in other business needs, including the cost-effectiveness of the controls and the innovation required to implement those controls.

An organization's current profile and target profile are very useful when conducting self-assessments and when communicating with various stakeholders across the organization or between your organization and partners or suppliers. These profiles ensure that everyone knows what the target is based on the target profile created and adopted by the organization, as well as the desired end state based on the target profile.

When developing your profiles, remember that there are 106 subcategories listed in the framework that you can choose from, but you do not need to use them all. For example, one of the subcategories under the Continuous Monitoring category within the Detect function is DE.CM-02. DE.CM-02 states, "The physical environment is monitored to find potentially adverse events."

If we were creating a profile for our company, AKYLADE, we would not select this outcome to include in our profiles. This is because AKYLADE operates as a remote first company and our entire infrastructure is built on serverless technology running across multiple cloud service provider networks and data centers around the globe. So, we don't have our own "physical" location with a large data center in it that we need to protect. In fact, at our offices, we don't have a ton of technology or a large network installed, and whether our employees are accessing our systems and services from our offices, from their homes, or from an airplane flying across the country, our systems and services are configured using the principles of zero-trust so that protecting our "physical environment" is not really a priority for us.

SUMMARY

Remember, the framework consists of three key components: the framework core, implementation tiers, and profile. The framework core is the framework's foundation and encompasses six key functions: Govern, Identify, Protect, Detect, Respond, and Recover. These functions are interconnected and form an operational culture that addresses the dynamic nature of cybersecurity risks.

The framework implementation tiers are used to categorize organizations based on their cybersecurity capability level, ranging from Tier 1 (partial) to Tier 4 (adaptive).

The framework profile, on the other hand, is used as a customizable tool that allows organizations to align the framework with their unique risk management objectives and prioritize their cybersecurity improvements. By understanding and leveraging these framework components, organizations can enhance their cyber resilience and effectively protect their critical assets and operations in today's highly connected digital world.

Keep in mind, though; it is imperative that you understand that the NIST Cybersecurity Framework is not a compliance requirement or regulation. The current NIST Cybersecurity Framework is considered to be voluntary. Therefore, it is important you are truthful when determining your organization's framework tier and creating its current profile and target profile. There is no right or wrong answer when developing these tiers within your organization, but instead, you need to ensure what you identify as your current state accurately reflects where you truly are and that your target state is where you truly plan to work across the entire organization.

CHAPTER SIX

SIX FUNCTIONS

Imagine a skilled orchestra where each musician plays a unique instrument, harmonizing their melodies and rhythms to create a captivating symphony. In the realm of cybersecurity, the NIST Cybersecurity Framework functions in a similar manner, with its six core functions working together in harmony to protect organizations against cyber threats and ensure the continuity of their operations. In this chapter, we will delve deeper into these six functions: Govern, Identify, Protect, Detect, Respond, and Recover. We will also explore their significance, outcomes, objectives, and the activities involved in each function's implementation.

The NIST Cybersecurity Framework provides a comprehensive roadmap for organizations to enhance their cybersecurity posture, and the six functions serve as the key pillars of this framework. Understanding these functions is crucial for organizations to establish effective cybersecurity practices, as they guide us in the governance of our organizational cybersecurity programs, identification of risks, implementation of protective measures, detection of potential incidents, response to attacks, and recovery from disruptions.

Each **function** plays a vital role, working in tandem to create a robust and resilient cybersecurity ecosystem. By gaining a deeper understanding of these functions, organizations can develop a more proactive and comprehensive approach to cybersecurity, aligning their strategies with industry best practices and leveraging existing standards and guidelines.

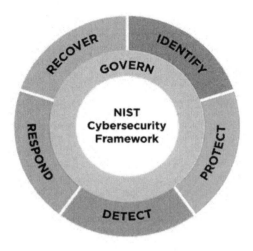

As we journey through the six functions – Govern, Identify, Protect, Detect, Respond, and Recover – we will uncover each one's specific actions and considerations. From conducting risk assessments and implementing access controls to deploying intrusion detection systems and developing incident response plans, each function offers a unique perspective and set of activities to strengthen an organization's cybersecurity defenses.

By following the guidance in this chapter, organizations can confidently navigate the complex landscape of cyber threats. The six functions will empower organizations to develop a holistic cybersecurity framework tailored to their specific needs, enabling them to establish governance, identify vulnerabilities, protect critical assets, detect potential breaches, respond effectively to incidents, and recover swiftly from disruptions. These six functions all work together to provide a high-level, strategic view of an organization's management lifecycle of any given cybersecurity risk.

Each function in the framework is divided into **categories**. Each **category** is labeled using a short code, signifying the function and the related category. For example, GV.RM represents that Risk Management strategy is a category under the Govern function.

Additionally, under each category, there are **subcategories** that add a two-digit number to the end of a given short code. For example, GV.RM-01 represents the subcategory specified as "Risk Management objectives are established and agreed to by organizational stakeholders" within the NIST Cybersecurity Framework.

There are 106 different subcategories spread amongst the six functions and 22 categories. When working as a cybersecurity consultant in the field, you can always reference the official framework documentation to determine which functions, categories, and subcategories would best align with your organizational needs.

Now, let's dive into the intricacies of the six functions and explore how they form the backbone of an organization's cyber resilience by fostering a secure environment in the face of evolving cyber threats and challenges.

GOVERN (GV)

The **Govern (GV)** function involves creating, communicating, implementing, and monitoring an organization's cybersecurity risk management strategy, expectations, and policy. This function provides expected outcomes that inform the organization as to what it may do to achieve and prioritize the outcomes of the other five functions and to ensure that those outcomes align with the organization's mission and stakeholder expectations.

In version 1.0 and 1.1 of the NIST Cybersecurity Framework, Govern was not its own function, but was instead part of the other five functions. In the latest version of the NIST Cybersecurity Framework, version 2.0, Govern was elevated to its own function, while still maintaining a lot of integration with the other five functions in the NIST Cybersecurity Framework.

The subcategories contained in the Govern function are considered critical for the incorporation of cybersecurity into an organizations' broader enterprise risk management (ERM) strategy. The Govern function serves as the baseline to establish, communicate, and monitor the organization's risk management strategy, expectations, and policies.

The Govern function contains six categories: Organizational Context; Risk Management Strategy; Roles, Responsibilities, and Authorities; Policy; Oversight; and Cybersecurity Supply Chain Risk Management.

Each Govern category contains two to ten different subcategories notated in the format of GV.xx-yy, where xx is the two-letter category code and yy represents the specific subcategory number.

GOVERN (GV) CATEGORIES AND SUBCATEGORIES

The **Organizational Context (GV.OC)** category of the Govern function involves understanding the circumstances surrounding the organization's risk management decisions, including the mission, stakeholder expectations, dependencies, and legal, regulatory, and contractual requirements.

Subcategories of the Organizational Context category include:

GV.OC-01 states that the organizational mission is understood and informs cybersecurity risk management. This means that the organization's mission has been clearly communicated and is considered when the organization makes risk management decisions.

GV.OC-02 states that the internal and external stakeholders are understood and that their needs and expectations regarding cybersecurity risk management are understood and considered.

GV.OC-03 states that the legal, regulatory, and contractual requirements regarding cybersecurity are understood and managed, including consideration of the organization's privacy and civil liberties obligations.

GV.OC-04 states that the critical objectives, capabilities, and services that external stakeholders depend on or expect from the organization are understood and communicated.

GV.OC-05 states that the outcomes, capabilities, and services that the organization depends on are understood and communicated.

The **Risk Management Strategy (GV.RM)** category of the Govern function ensures the organization's priorities, constraints, risk tolerance and appetite statements, and assumptions are established, communicated, and used to support operational risk decisions.

Subcategories of the Risk Management Strategy category include:

GV.RM-01 states that the risk management objectives are established and agreed to by organizational stakeholders.

GV.RM-02 states that the risk appetite and risk tolerance statements are established, communicated, and maintained.

GV.RM-03 states that the cybersecurity risk management activities and outcomes are included in enterprise risk management processes.

GV.RM-04 states that the strategic direction that describes appropriate risk response options is established and communicated.

GV.RM-05 states that the lines of communication across the organization are established for cybersecurity risks, including risks from suppliers and other third parties.

GV.RM-06 states that a standardized method for calculating, documenting, categorizing, and prioritizing cybersecurity risks is established and communicated.

GV.RM-07 states that strategic opportunities (i.e., positive risks) are characterized and are included in organizational cybersecurity risk discussions.

The **Roles, Responsibilities, and Authorities (GV.RR)** category of the Govern function establishes cybersecurity roles, responsibilities, and authorities to foster accountability, performance assessment, and continuous improvement are established and communicated.

Subcategories of the Roles, Responsibilities, and Authorities category include:

GV.RR-01 states that organizational leadership is responsible and accountable for cybersecurity risk and fosters a culture that is risk-aware, ethical, and continually improving.

GV.RR-02 states that the roles, responsibilities, and authorities related to cybersecurity risk management are established, communicated, understood, and enforced.

GV.RR-03 states that the adequate resources are allocated commensurate with the cybersecurity risk strategy, roles, responsibilities, and policies.

GV.RR-04 states that cybersecurity is included in human resources practices.

The **Policy (GV.PO)** category of the Govern function ensures that the organization's cybersecurity policy is established, communicated, and enforced.

Subcategories of the Policy category include:

GV.PO-01 states that the policy for managing cybersecurity risks is established based on organizational context, cybersecurity strategy, and priorities and is communicated and enforced.

GV.PO-02 states that the policy for managing cybersecurity risks is reviewed, updated, communicated, and enforced to reflect changes in requirements, threats, technology, and organizational mission.

The **Oversight (GV.OV)** category of the Govern function ensures that the results of organization-wide cybersecurity risk management activities and performance are used to inform, improve, and adjust the risk management strategy.

Subcategories of the Oversight category include:

GV.OV-01 states that cybersecurity risk management strategy outcomes are reviewed to inform and adjust strategy and direction.

GV.OV-02 states that the cybersecurity risk management strategy is reviewed and adjusted to ensure coverage of organizational requirements and risks.

GV.OV-03 states that organizational cybersecurity risk management performance is evaluated and reviewed for adjustments needed.

The **Cybersecurity Supply Chain Risk Management (GV.SC)** category of the Govern function ensures that the cyber supply chain risk management processes are identified, established, managed, monitored, and improved by organizational stakeholders. In NIST Cybersecurity Framework version 2.0 there is an increased focus on supply chain risk management as it has become an increasingly common vector for cyber attacks in recent years.

Subcategories of the Cybersecurity Supply Chain Risk Management category include:

GV.SC-01 states that a cybersecurity supply chain risk management program, strategy, objectives, policies, and processes are established and agreed to by organizational stakeholders.

GV.SC-02 states that the cybersecurity roles and responsibilities for suppliers, customers, and partners are established, communicated, and coordinated internally and externally.

GV.SC-03 states that cybersecurity supply chain risk management is integrated into cybersecurity and enterprise risk management, risk assessment, and improvement processes.

GV.SC-04 states that the suppliers are known and prioritized by criticality.

GV.SC-05 states that the requirements to address cybersecurity risks in supply chains are established, prioritized, and integrated into contracts and other types of agreements with suppliers and other relevant third parties.

GV.SC-06 states that planning and due diligence are performed to reduce risks before entering into formal supplier or other third-party relationships.

GV.SC-07 states that the risks posed by a supplier, their products and services, and other third parties are understood, recorded, prioritized, assessed, responded to, and monitored over the course of the relationship.

GV.SC-08 states that relevant suppliers and other third parties are included in incident planning, response, and recovery activities.

GV.SC-09 states that the supply chain security practices are integrated into cybersecurity and enterprise risk management programs, and their performance is monitored throughout the technology product and service life cycle.

GV.SC-10 states that cybersecurity supply chain risk management plans include provisions for activities that occur after the conclusion of a partnership or service agreement.

IDENTIFY (ID)

The **Identify (ID)** function involves developing an organizational understanding of cybersecurity risks to an organization's assets, including its data, hardware, software, systems, facilities, services, people, suppliers, and capabilities. This function helps to prioritize and align organizational resources with its risk management strategy and its cybersecurity efforts based on the organization's mission.

The subcategories contained in the Identify function are considered foundational for the effective use of the NIST Cybersecurity Framework. The Identify function is used to understand the business within the environmental and organizational context, determine which resources support critical functions, and recognize the related cybersecurity risks that might affect the organization. This enables your organization to focus and prioritize its cybersecurity defense efforts while being consistent with its overall risk management strategy and the organization's mission. Additionally, the Identify function is used to identify improvement opportunities for the organization's policies, plans, processes, procedures, and practices in order to inform its efforts across all six of the NIST Cybersecurity Framework functions.

The Identify function contains three categories: Asset Management; Risk Assessment; and Improvement.

Each Identify category contains four to ten different subcategories notated in the format of ID.xx-yy, where xx is the two-letter category code and yy represents the specific subcategory number. Note that the yy number may not always be sequential because the NIST Cybersecurity Framework adds and removes subcategories in each new version of the framework without renumbering each subcategory.

IDENTIFY (ID) CATEGORIES AND SUBCATEGORIES

The **Asset Management (ID.AM)** category of the Identify function involves the identification of assets, data, hardware, software, systems, facilities, services, and people that enable the organization to achieve its business purposes. These assets must be identified and managed consistently with their relative importance to organizational objectives and risk strategy.

Subcategories of the Asset Management category include:

ID.AM-01 states that inventories of hardware managed by the organization are maintained.

ID.AM-02 states that inventories of software, services, and systems managed by the organization are maintained.

ID.AM-03 states that representations of the organization's authorized network communication and internal and external network data flows are maintained.

ID.AM-04 states that inventories of services provided by suppliers are maintained.

ID.AM-05 states that assets are prioritized based on classification, criticality, resources, and impact on the mission.

ID.AM-07 states that inventories of data and corresponding metadata for designated data types are maintained.

ID.AM-08 states that systems, hardware, software, services, and data are managed throughout their life cycles.

The **Risk Assessment (ID.RA)** category of the Identify function ensures that the organization understands the cybersecurity risk to its organizational operations (including mission, functions, image, or reputation), organizational assets, and individuals.

Subcategories of the Risk Management category include:

ID.RA-01 states that vulnerabilities in assets are identified, validated, and recorded.

ID.RA-02 states that cyber threat intelligence is received from information sharing forums and sources.

ID.RA-03 states that internal and external threats to the organization are identified and recorded.

ID.RA-04 states that the potential impacts and likelihoods of threats exploiting vulnerabilities are identified and recorded.

ID.RA-05 states that the threats, vulnerabilities, likelihoods, and impacts are used to understand inherent risk and inform risk response prioritization.

ID.RA-06 states that risk responses are chosen, prioritized, planned, tracked, and communicated.

ID.RA-07 states that changes and exceptions are managed, assessed for risk impact, recorded, and tracked.

ID.RA-08 states that the processes for receiving, analyzing, and responding to vulnerability disclosures are established.

ID.RA-09 states that the authenticity and integrity of hardware and software are assessed prior to acquisition and use.

ID.RA-10 states that critical suppliers are assessed prior to acquisition.

The **Improvement (ID.IM)** category of the Identify function that is focused on improvements to organizational cybersecurity risk management processes, procedures, and activities that have been identified across all of the NIST Cybersecurity Framework functions.

The subcategories of the Improvement category include:

ID.IM-01 states that improvements are identified from evaluations.

ID.IM-02 states that improvements are identified from security tests and exercises, including those done in coordination with suppliers and relevant third parties.

ID.IM-03 states that improvements are identified from execution of operational processes, procedures, and activities.

ID.IM-04 states that incident response plans and other cybersecurity plans that affect operations are established, communicated, maintained, and improved.

PROTECT (PR)

The **Protect (PR)** function is used by organizations to develop and implement safeguards to ensure the delivery of critical services and the protection of physical and digital assets against cyber threats. The Protect function supports the organization's ability to limit or contain the impact of a potential cybersecurity event or incident.

The Protect function contains five categories: Identity Management, Authentication and Access Control; Awareness and Training; Data Security; Platform Security; and Technology Infrastructure Resilience.

Each Protect category contains two to six different subcategories notated in the format of PR.xx-yy, where xx is the two-letter category code and yy represents the specific subcategory number. Note that the yy number may not always be sequential because the NIST Cybersecurity Framework adds and removes subcategories in each new version of the framework without renumbering each subcategory.

PROTECT (PR) CATEGORIES AND SUBCATEGORIES

The **Identity Management, Authentication, and Access Control (PR.AA)** category of the Protect function is used to implement effective mechanisms for the management of user identities, ensuring proper authentication processes, and controlling access to systems and resources to prevent unauthorized activities. This control focuses on the access to physical and logical assets and associated facilities being limited to authorized users, processes, services, hardware, and devices. Its management should also be consistent with the assessed risk of unauthorized access to authorized activities and transactions.

Subcategories of the Identity Management, Authentication, and Access Control category include:

PR.AA-01 states that identities and credentials for authorized users, services, and hardware are managed by the organization.

PR.AA-02 states that identities are proofed and bound to credentials based on the context of interactions.

PR.AA-03 states that users, services, and hardware are authenticated.

PR.AA-04 states that the identity assertions are protected, conveyed, and verified.

PR.AA-05: Access permissions, entitlements, and authorizations are defined in a policy, managed, enforced, and reviewed, and incorporate the principles of least privilege and separation of duties.

PR.AA-06: Physical access to assets is managed, monitored, and enforced commensurate with risk.

The **Awareness and Training (PR.AT)** category of the Protect function emphasizes the importance of educating and raising awareness among personnel about cybersecurity risks, threats, and best practices to foster a security-conscious culture and enhance the organization's overall cybersecurity posture. The organization's personnel and partners should be provided cybersecurity awareness education and trained to perform their cybersecurity-related duties and responsibilities consistent with the organization's related policies, procedures, and agreements.

Subcategories of the Awareness and Training category include:

PR.AT-01 states that personnel are provided with awareness and training so that they possess the knowledge and skills to perform general tasks with cybersecurity risks in mind.

PR.AT-02 states that individuals in specialized roles are provided with awareness and training so that they possess the knowledge and skills to perform relevant tasks with cybersecurity risks in mind.

The **Data Security (PR.DS)** category of the Protect function focuses on protecting the confidentiality, integrity, and availability of sensitive data within an organization's systems and networks, ensuring appropriate safeguards are in place to mitigate data breaches and unauthorized access. It involves establishing and implementing data protection measures, controls, and procedures to secure data at rest, in transit, and during processing to maintain the organization's data integrity and prevent data loss or compromise.

Subcategories of the Data Security category include:

PR.DS-01 states that the confidentiality, integrity, and availability of data-at-rest are protected.

PR.DS-02 states that the confidentiality, integrity, and availability of data-in-transit are protected.

PR.DS-10 states that the confidentiality, integrity, and availability of data-in-use are protected.

PR.DS-11 states that the backups of data are created, protected, maintained, and tested.

The **Platform Security (PR.PS)** category of the Protect function focuses on the organization's consistent management of its risk strategy to protect its confidentiality, integrity, and availability of its hardware, software, and services of both its physical and virtual platforms.

Subcategories of the Platform Security category include:

PR.PS-01 states that configuration management practices are established and applied.

PR.PS-02 states that the software is maintained, replaced, and removed commensurate with risk.

PR.PS-03 states that the hardware is maintained, replaced, and removed commensurate with risk.

PR.PS-04 states that log records are generated and made available for continuous monitoring.

PR.PS-05 states that the installation and execution of unauthorized software are prevented.

PR.PS-06 states that secure software development practices are integrated, and their performance is monitored throughout the software development life cycle.

The **Technology Infrastructure Resilience (PR.IR)** category of the Protect function ensures that the security architectures are managed in alignment with the organization's risk strategy to protect the confidentiality, integrity, availability of its assets while maintaining its organization resilience.

Subcategories of the Technology Infrastructure Resilience category include:

PR.IR-01 states that networks and environments are protected from unauthorized logical access.

PR.IR-02 states that the organization's technology assets are protected from environmental threats.

PR.IR-03 states that mechanisms are implemented to achieve resilience requirements in normal and adverse situations.

PR.IR-04 states that adequate resource capacity to ensure availability is maintained.

DETECT (DE)

The **Detect (DE)** function is used by organizations to develop and implement appropriate activities to identify the occurrence of a cybersecurity event through timely discovery and analysis of anomalies, indicators of compromise, and other potentially adverse events. The Detect function enables organizations to successfully detect potential cybersecurity attacks and compromises in order to rapidly begin incident response and recovery efforts.

The Detect function contains two categories: Continuous Monitoring and Adverse Event Analysis.

Each Detect category contains five or six different subcategories notated in the format of DE.xx-yy, where xx is the two-letter category code and yy represents the specific subcategory number. Note that the yy number may not always be sequential because the NIST Cybersecurity Framework adds and removes subcategories in each new version of the framework without renumbering each subcategory.

DETECT (DE) CATEGORY AND SUBCATEGORIES:

The **Continuous Monitoring (DE.CM)** category of the Detect function ensures that the information system and assets are continually monitored to identify anomalies, indicators of compromise, and potentially adverse cybersecurity events.

Subcategories of the Continuous Monitoring category include:

DE.CM-01 states that networks and network services are monitored to find potentially adverse events.

DE.CM-02 states that the physical environment is monitored to find potentially adverse events.

DE.CM-03 states that personnel activity and technology usage are monitored to find potentially adverse events.

DE.CM-06 states that external service provider activities and services are monitored to find potentially adverse events.

DE.CM-09 states that the computing hardware and software, runtime environments, and their data are monitored to find potentially adverse events.

The **Adverse Event Analysis (DE.AE)** category of the Detect function focuses on the analysis of anomalies, indicators of compromise, and other potentially adverse cyber events to characterize the events and detect cybersecurity incidents.

Subcategories of the Adverse Event Analysis category include:

DE.AE-02 states that potentially adverse events are analyzed to better understand associated activities.

DE.AE-03 states that the information is correlated from multiple sources.

DE.AE-04 states that the estimated impact and scope of adverse events are understood.

DE.AE-06 states that information on adverse events is provided to authorized staff and tools.

DE.AE-07 states that cyber threat intelligence and other contextual information are integrated into the analysis.

DE.AE-08 states that incidents are declared when adverse events meet the defined incident criteria.

RESPOND (RS)

The **Respond (RS)** function is used by organizations to develop and implement appropriate activities to perform actions regarding a detected cybersecurity incident. The Respond function supports the ability of the organization to contain the impact of a potential cybersecurity incident.

The Respond function contains four categories: Incident Management; Incident Analysis; Incident Response Reporting and Communication; and Incident Mitigation.

Each Respond category contains two to five different subcategories notated in the format of RS.xx-yy, where xx is the two-letter category code and yy represents the specific subcategory number. Note that the yy number may not always be sequential because the NIST Cybersecurity Framework adds and removes subcategories in each new version of the framework without renumbering each subcategory.

RESPOND (RS) CATEGORIES AND SUBCATEGORIES:

The **Incident Management (RS.MA)** category of the Respond focuses on how an organization manages their response efforts for detected cybersecurity incidents. This includes executing and maintaining response plans, processes, and procedures when responding to anomalies and potential adverse cyber events.

Subcategories of the Incident Management category include:

RS.MA-01 states that the incident response plan is executed in coordination with relevant third parties once an incident is declared.

RS.MA-02 states that incident reports are triaged and validated.

RS.MA-03 states that incidents are categorized and prioritized.

RS.MA-04 states that incidents are escalated or elevated as needed.

RS.MA-05 states that the criteria for initiating incident recovery are applied.

The **Incident Analysis (RS.AN)** category of the Respond function ensures that proper analysis is conducted to respond effectively and to support the organization's forthcoming recovery activities. The organization should conduct investigations that implement effective response actions, support forensic activities, and prepare for their future recovery efforts.

Subcategories of the Incident Analysis category include:

RS.AN-03 states that the analysis is performed to establish what has taken place during an incident and the root cause of the incident.

RS.AN-06 states that the actions performed during an investigation are recorded, and the records' integrity and provenance are preserved.

RS.AN-07 states that incident data and metadata are collected, and their integrity and provenance are preserved.

RS.AN-08 states that an incident's magnitude is estimated and validated.

The **Incident Response Reporting and Communication (RS.CO)** category of the Respond function focuses on ensuring that all response activities are coordinated with internal and external stakeholders as required by laws, regulations, or policies.

Subcategories of the Incident Response Reporting and Communication category include:

RS.CO-02 states that internal and external stakeholders are notified of incidents.

RS.CO-03 states that the information is shared with designated internal and external stakeholders.

The **Incident Mitigation (RS.MI)** category of the Respond function ensures that activities are performed to prevent the expansion of an event, mitigate its effects, and resolve the incident.

Subcategories of the Incident Mitigation category include:

RS.MI-01 states that incidents are contained.

RS.MI-02 states that incidents are eradicated.

RECOVER (RC)

The **Recover (RC)** function helps an organization develop and implement appropriate activities to maintain resilience plans and restore any capabilities or services that were impaired due to a cybersecurity incident. The Recover function focuses on supporting an organization's timely restoration to normal operations which can reduce the impact of a potentially adverse cyber event.

The Recover function contains two categories: Incident Recovery Plan Execution and Incident Recovery Communication.

Each Recover category contains two or six different subcategories notated in the format of RC.xx-yy, where xx is the two-letter category code and yy represents the specific subcategory number. Note that the yy number may not always be sequential because the NIST

Cybersecurity Framework adds and removes subcategories in each new version of the framework without renumbering each subcategory.

RECOVER (RC) CATEGORIES AND SUBCATEGORIES:

The **Incident Recovery Plan Execution (RC.RP)** category of the Recover function focuses on the execution and maintenance of the recovery processes and procedures to ensure that an organization's systems and services have been restored after a cybersecurity incident to full operational availability.

Subcategories of the Incident Recovery Plan Execution category include:

RC.RP-01 states that the recovery portion of the incident response plan is executed once initiated from the incident response process.

RC.RP-02 states that recovery actions are selected, scoped, prioritized, and performed.

RC.RP-03 states that the integrity of backups and other restoration assets is verified before using them for restoration.

RC.RP-04 states that critical mission functions and cybersecurity risk management are considered to establish post-incident operational norms.

RC.RP-05 states that the integrity of restored assets is verified, systems and services are restored, and normal operating status is confirmed.

RC.RP-06 states that the end of incident recovery is declared based on criteria, and incident-related documentation is completed.

The **Incident Recovery Communication (RC.CO)** category of the Recover function ensures that all restoration activities are coordinated with internal and external parties, such as with their coordinating centers, Internet Service Providers, owners of attacking systems, victims, other cybersecurity incident response teams, and vendors, as appropriate.

Subcategories of the Incident Recovery Communication category include:

RC.CO-03 states that recovery activities and progress in restoring operational capabilities are communicated to designated internal and external stakeholders.

RC.CO-04 states that public updates on incident recovery are shared using approved methods and messaging.

SUMMARY

This chapter covered all six functions, 22 categories, and 106 subcategories of the NIST Cybersecurity Framework version 2.0. It is important that you can relate a given subcategory to the appropriate category and the category to the appropriate function.

For the exam, you will not be asked to expand a given short code from memory, such as DE.AE-04. However, you may receive questions on the exam that ask you to identify the function or category to which DE.AE-04 is linked. In this case, it would be sufficient for you to remember that DE is the Detect function and AE is the Adverse Event Analysis category.

Just to be clear, for the exam you will be expected to know the six functions and the 22 categories, but specific questions about the 106 subcategories are beyond the scope of the exam.

When you are working in the field as a cybersecurity consultant, you can always carry a copy of the NIST Cybersecurity Framework version 2.0 to reference the individual subcategories, as needed. As of version 2.0 of the framework, NIST now provides an online tool that includes all the functions, categories, and subcategories with a link to the informative references that include specific security controls for each subcategory, which you can find at https://www.nist.gov/cyberframework.

It is important to remember that within the 22 categories are specific subcategories, such as inventorying physical devices and systems, mapping organizational communication and data flows, and establishing cybersecurity roles and responsibilities. These 106 subcategories enable organizations to prioritize resources, assess risks, and develop effective risk management strategies.

As you work in the field, you will find that you don't need to use all 106 subcategories with every organization. Instead, it is more important that you review them all during your engagements and select the appropriate subcategories that best align with organizational objectives and overall risk management strategies being pursued.

CHAPTER SEVEN

CONTROLS AND OUTCOMES

Imagine you're about to get into your car and drive to work. Before you even leave your driveway, there is an essential step that you must always take: putting on your seatbelt. But why?

Well, in most countries around the world, there are laws that dictate you must wear a seatbelt when driving a car. If you don't wear your seatbelt and a police officer pulls you over, you may receive a ticket or a fine for not following this regulation.

But, seatbelt laws were not passed by our governments to simply collect additional revenue through fines. Instead, these regulations were created to save people's lives.

In the event of a sudden stop or collision, your seatbelt should keep you safely in your seat instead of allowing forward momentum to throw you through your windshield.

In this chapter, we will explore the different controls and outcomes used within the NIST Cybersecurity Framework. We will observe that, just like the seatbelt in your car, controls and outcomes in cybersecurity serve as vital safeguards that help protect your systems, assets, and data from potential cyber threats.

So, fasten your cybersecurity seatbelt as we explore the world of controls and outcomes, discovering the key practices that will keep your organization secure when operating on the information superhighway.

CONTROLS

If you have used other cybersecurity frameworks, like the NIST Risk Management Framework (RMF), the Center of Internet Security Critical Security Controls (CIS CSC), COBIT 5, or the ISO/IEC 27001 series, you are probably used to dealing with controls.

In cybersecurity, **controls** refer to specific measures, practices, or safeguards that organizations implement to manage and mitigate cybersecurity risks. These controls are tangible actions or mechanisms that help prevent, detect, respond to, and recover from cybersecurity incidents. They can include technical solutions, policies, procedures, training programs, and other security measures designed to protect systems, assets, and data.

For example, one control for increasing the security of an organization's authentication system might require all users to utilize a long, complex password containing at least 16 characters and a mixture of uppercase, lowercase, numeric, and special characters. If the organization has a higher security system they wish to protect, they could instead opt to remove password-based logins from their system completely and migrate to a multi-factor authentication system based on a smart card and PIN for user logins.

The idea with control is that the organization wants to apply countermeasures that make it more difficult for a threat actor to compromise the organization's systems, assets, and data. These controls act as your organization's safety mechanism, mitigating risks and providing a level of assurance against potential cybersecurity incidents from occurring.

Depending on the framework used, there will be a list of controls that can be selected to ensure the organization complies with that given framework. In most frameworks, controls are a requirement that must be followed for all organizations. These are known as prescriptive controls, and they are specific and detailed control measures that provide explicit instructions or requirements on how to implement a security measure or safeguard in a standardized and structured manner.

Unlike many prescriptive frameworks, the NIST Cybersecurity Framework doesn't utilize these prescriptive or mandatory controls. Instead, this framework focuses on using outcomes to achieve higher levels of cybersecurity tailored for specific organizational needs.

VOLUNTARY NATURE OF THE NIST CYBERSECURITY FRAMEWORK

The NIST Cybersecurity Framework is completely voluntary for companies and organizations to use. As such, it was developed with no specific prescriptive controls. Organizations using the framework are given wide latitude and freedom to perform the functions and activities in any way that makes sense in order to achieve their desired outcomes. These outcomes are set forth in the six core functions, 22 categories, and 106 subcategories.

When many first hear that the NIST Cybersecurity Framework does not have specific prescriptive controls, they immediately believe this makes it less secure. This is an incorrect assumption, though. In fact, one of the biggest issues with more prescriptive frameworks is that they set forth requirements that MUST be complied with due to their prescriptive nature, even though the requirements add no additional security benefits.

A great example of this is the password complexity requirement mentioned earlier. Many people believe that long, complex passwords are more secure because they increase the time required for an attacker to guess or brute force that password. In theory, this would be a true statement, but that's not how it works in the real world.

Studies have shown time and time again that long, complex passwords are often less secure than a less complex password might be. Why would that be?

Well, it comes down to human nature. Humans have a hard time remembering a twenty-character complex password with lots of different letters, numbers, and symbols. For example, a very long, strong, and complex password might be a 32-character randomly generated password like @Q*Qi6sTAYV4TkU4oTrhs3s-XPKhMo42. Now, even though this is a very long, strong, and complex password, most people will be unable to remember it. So, they simply write it down or type it into a virtual sticky note on their computer.

So, if an attacker compromises the person's computer using an exploit or uses social engineering to gain access to the machine, they could now locate the password stored in the virtual sticky note. During on-site penetration tests, we have often observed physical paper notes located underneath an employee's keyboard with their long, strong password written there for anyone to see. Therefore, longer and more complex passwords become less secure when applied to real-world conditions.

Instead, it would be better for the organization to implement a good multi-factor authentication. But, if the prescriptive control in another framework states that you must use a long, complex password, then you can't enable multi-factor authentication because your organization would not comply with the password length and complexity control in that framework.

This is where the true power of the NIST Cybersecurity Framework comes into play because it focuses on outcomes, not prescriptive controls. How an organization will achieve that outcome is left up to the organization to determine.

If you need some guidance for finding controls to reach your outcomes, you can look at the informative references which provide the compensating controls that other frameworks require, such as Security and Privacy Controls for Federal Information Systems and Organizations (NIST SP 800-53 Rev. 5), Center for Internet Security Critical Security Controls (CIS CSC), Control Objectives for Information and Related

Technology (COBIT 5), Security for Industrial Automation and Control Systems (ISA 62443-3-2:2009) and (ISA 62443-3-3:2013), and Information Security Management System (ISO/IEC 27001:32013) Risk Management Strategy.

As an organization begins to develop its profiles and lists the different outcomes it wants to achieve, it can then choose and implement its own internal controls, or it may opt to use the controls from another framework. But the key is the organization itself oversees its cybersecurity risks. It has the power to choose the proper controls from any source needed to deliver the outcomes chosen in the organizational target profile.

OUTCOMES

An **outcome** refers to the desired result or objective that an organization aims to achieve through implementing cybersecurity controls and practices, focusing on those measures' overall effectiveness and impact. Simply put, an outcome is a change or result the organization expects to observe from a given process or action.

In the NIST Cybersecurity Framework, outcomes are specific to each category and subcategory within the framework. They provide guidance on the intended goals and results that an organization should strive for when implementing controls related to a particular aspect of cybersecurity.

The one thing people struggle with when they are new to the NIST Cybersecurity Framework is that each outcome is written as a requirement. For example, we find the category PR.DS under the Protect function to signify the Data Security category. PR.DS-11 states that "[b]ackups of data are created, protected, maintained, and tested." This sounds extremely broad and generic to newcomers to the NIST Cybersecurity Framework, causing them frustration.

To increase their productivity, many cyber resiliency professionals have learned to take these outcomes and rewrite them into testable pieces or questions before they use them in evaluating how well an organization delivers results for each chosen outcome.

For example, if PR.DS-11 is going to be broken down into smaller, testable items, we might have four separate and specific requirements:

1. Backups of data are created.
2. Backups of data are protected.
3. Backups of data are maintained.
4. Backups of data are tested.

For each of these requirements, the cyber resiliency professional might then convert these specific requirements into testable questions that can be used during their internal assessment of an organization to determine if they are following best practices and recommendations for creating, protecting, maintaining, and testing the system backups.

It is important to remember that the NIST Cybersecurity Framework is not a checklist because a checklist is a list of required items, things to do, or points to be considered. In fact, NIST is very clear about this point in their guidance for the implementation of the NIST Cybersecurity Framework. The framework was never designed to be a mandatory compliance framework or used as a prescriptive checklist of controls.

Consider another example from within the Detect function. In the Detect function, there is a category called Continuous Monitoring (DE.CM). Under this category, there are five subcategories that can help define the specific outcomes that need to be achieved in order to state that the organization's continuous monitoring is satisfactory.

For instance, DE.CM-02 states that "the physical environment is monitored to find potential adverse events." This activity, though, is being performed to achieve an outcome and not to meet a prescriptive control in the framework. The organization can determine what actions will meet their needs regarding exactly how they wish to monitor their physical environment.

Some questions for an organization that might have physical environment related needs might include:

Will the organization utilize security cameras along their perimeter to monitor their physical environment?

Will the organization hire security guards to roam throughout their building 24 hours a day to actively monitor their facilities?

Will the organization install an eight-foot tall, barbed wire fence with motion sensors to continuously monitor their physical environment?

Will the organization conduct physical penetration tests to determine if their physical environment is being properly secured?

All these questions are left up to the organization to determine what best meets their needs because the NIST Cybersecurity Framework is not prescriptive in nature.

For example, if the organization is trying to protect its data center that contains credit card processing systems, it may fall under the requirements of the PCI-DSS contractual obligations. This may require them to perform physical penetration tests using an external consultant who is certified by the Payment Card Industry to perform that type of work.

If this organization is also a healthcare provider, it may fall under the Health Insurance Portability and Accountability Act of 1996 (HIPAA) requirements, as well.

Both the PCI-DSS and HIPAA requirements will have certain controls that must be used, but the organization can combine all of their required controls across all regulations and create a singular action plan using the NIST Cybersecurity Framework, as well.

Regarding this type of physical monitoring, the NIST Cybersecurity Framework only requires that you achieve your organizational desired outcome. In this case, your organization may set its desired outcome as the ability to identify and remediate 98% of known physical intrusion vulnerabilities within 90 days of discovery. To verify that is being done, the organization may need to perform physical penetration tests every 90 days to determine what vulnerabilities have been patched or mitigated and which still remain. Alternatively, your organization may determine that it wants to set its desired outcome as only 90% of known physical intrusion vulnerabilities within a 120-day window. Either is fine as far as the NIST Cybersecurity Framework is concerned because it gives the organization the flexibility to determine its desired outcomes instead of having to meet more traditional, prescriptive framework standards and targets. This is the flexibility the NIST Cybersecurity Framework provides an organization over a traditional, more prescriptive framework.

Remember, outcomes serve as measurable indicators of the effectiveness and maturity of an organization's cybersecurity practices. They provide a tangible way to assess progress and determine if the desired objectives are being met. By focusing on outcomes, organizations can align their efforts with specific goals, continuously monitor their performance, and make informed decisions to improve their cybersecurity posture.

It is important to note that outcomes are not rigid requirements, but rather flexible guidelines that can be adapted and tailored to each organization's unique needs and priorities. This allows for flexibility in implementation while ensuring that the desired security objectives are being addressed.

The NIST Cybersecurity Framework provides you with the latitude necessary to adjust the framework to meet your organization's needs. Many people find this breadth and flexibility frustrating when they first begin using the framework because they may want to be clearly told what they need to do and which controls to implement. The problem with this rigid approach, however, is that it doesn't scale very well. Prescriptive frameworks tend to be less relevant in the long term because systems and the technology controls needed to secure those systems change at an alarming rate these days.

INFORMATIVE REFERENCES

If the organization wants to combine the NIST Cybersecurity Framework with some other more prescriptive, control-based frameworks, it certainly can. In fact, all of the subcategories listed in Appendix A of the NIST Cybersecurity Framework version 2.0 are linked to the "CSF 2.0

Informative Reference in the Core" tool at
https://www.nist.gov/informative-references.

This tool contains a searchable database that contains all six functions, 22 categories, and 106 subcategories. For each subcategory the tool provides **implementation examples** along with identification of the control inside of the informative reference.

An **informative reference** is a specific section of standards, guidelines, and practices that illustrate a method to achieve the outcomes associated with each category and subcategory. Informative references include controls from CIS CSC, COBIT 5, ISA 62443-3-3:2013, ISO/IEC 27001:2013, and NIST SP 800-53, as well as many other frameworks and control sets.

For example, in the "CSF 2.0 Information Reference in the Core" tool, you can locate the subcategory PR.DS-01 that states "confidentiality, integrity, and availability of data-at-rest are protected." For this subcategory, there are 26 informative references that link back to other frameworks, which you can use as controls within your organization.

Additionally, if the organization follows the ISO/IEC 27001:2013 standard, then you can select that line item as the control you wish to utilize to ensure data-at-rest is protected. In this case, that would be ISO/IEC 27001:2013 A.8.2.3.

The NIST Cybersecurity Framework is simply acting as an index to other established controls, in this case, so to determine what the control requires, you would need to look it up in the ISO/IEC 27001:2013 under A.8.2.3 which covers the handling of assets, including the storage of those assets.

OTHER FRAMEWORKS

By looking at the informative references in the "CSF 2.0 Informative Reference in the Core" tool, you will see many different frameworks being referenced, including many of the following:

- International Organization for Standardization (ISO)/International Electrotechnical Commission (IEC) 27001 and 27002
- National Institute of Standards and Technology (NIST) Special Publications (SP 800-37, SP 800-53, SP 800-171, SP 800-218, and SP 800-221A)
- Center for Internet Security Critical Security Controls (CIS CSC)
- Control Objectives for Information and Related Technology (COBIT 5)
- Cyber Risk Institute (CRI) Profile
- Information Technology Infrastructure Library (ITIL)
- Payment Card Industry Data Security Standard (PCI-DSS)
- Health Insurance Portability and Accountability Act (HIPAA)
- North American Electric Reliability Corporation (NERC) Critical Infrastructure Protection (CIP) Standards
- Federal Risk and Authorization Management Program (FedRAMP)
- Open Web Application Security Project (OWASP)
- Cloud Security Alliance (CSA) Security, Trust, Assurance, and Risk (STAR) Registry

For the certification exams, you are not required to know any of these frameworks in-depth, but you should have a basic understanding of the types of information they contain and that they can be used as sources for controls to achieve the organization's desired outcomes within the NIST Cybersecurity Framework.

INTERNATIONAL ORGANIZATION FOR STANDARDIZATION (ISO)/INTERNATIONAL ELECTROTECHNICAL COMMISSION (IEC) 27001 AND 27002

The International Organization for Standardization (ISO)/International Electrotechnical Commission (IEC) 27001 and 27002 framework is a set of international standards that provide guidelines and best practices for establishing, implementing, maintaining, and continually improving an information security management system (ISMS). ISO/IEC 27001 specifies the requirements for establishing and maintaining an ISMS, while ISO/IEC 27002 provides a comprehensive set of controls and implementation guidance for information security management.

ISO/IEC 27001 focuses on the management aspects of information security, emphasizing the need for a systematic approach to identify, assess, and manage information security risks within an organization. It provides a framework for establishing policies, processes, and procedures to ensure information assets' confidentiality, integrity, and availability. The controls outlined in ISO/IEC 27001 cover a wide range of areas, including information security policies; organizational security; asset management; human resource security; physical and environmental security; communications and operations management; access control; information systems acquisition; development and maintenance; and incident management, among others.

Organizations can leverage the ISO/IEC 27001 and 27002 frameworks to enhance their cybersecurity posture when used in conjunction with the NIST Cybersecurity Framework. The NIST Cybersecurity Framework provides a flexible and risk-based approach to managing and mitigating cyber risks, while ISO/IEC 27001 and 27002 offer detailed controls and implementation guidance. By aligning the two frameworks, organizations can benefit from the comprehensive control framework of ISO/IEC 27001 and 27002 while leveraging the risk management and organizational framework provided by the NIST Cybersecurity Framework.

NIST SPECIAL PUBLICATIONS
(SP 800-37, SP 800-53, SP 800-171, SP 800-218, and SP 800-221A)

The National Institute of Standards and Technology (NIST) Special Publications (SP 800-37, SP 800-53, SP 800-171, SP 800-218, and SP 800-221A) are a collection of guidelines, standards, and procedures developed by NIST to assist organizations in managing and enhancing their cybersecurity practices. These publications provide comprehensive guidance on various aspects of cybersecurity, including risk management, security controls, and security assessment and authorization.

SP 800-37, known as the "Risk Management Framework for Information Systems and Organizations," provides a structured and flexible approach to managing cybersecurity risks. The Risk Management Framework (RMF) outlines the process for selecting and implementing security controls based on risk assessments, continuous monitoring, and ongoing authorization. SP 800-37 helps organizations integrate risk management into their overall cybersecurity program and ensures that security controls are effectively implemented and maintained throughout the system's lifecycle.

SP 800-53, also known as the "Security and Privacy Controls for Information Systems and Organizations," outlines a set of security controls that can be implemented to protect information systems' confidentiality, integrity, and availability. It provides a catalog of control families, such as access control, incident response, and system and information integrity, each containing specific controls and implementation guidance.

SP 800-171, titled "Protecting Controlled Unclassified Information in Nonfederal Systems and Organizations," focuses on safeguarding sensitive information that is not classified but still requires protection. It provides a set of security requirements that nonfederal organizations must meet when handling controlled unclassified information (CUI) on behalf of the federal government. These requirements cover areas such as access control, media protection, incident response, and system and communications protection.

SP 800-218, known as the "Secure Software Development Framework (SSDF) Version 1.1," provides a set of high-level secure software development practices that can be integrated into an organization's software development lifecycle. By following the practices in this guide, organizations can reduce the number of vulnerabilities in the software they create, mitigate potential exploitation of undetected or unaddressed vulnerabilities, and quickly address the root causes of vulnerabilities to prevent future exploitation.

SP 800-221A, known as the "Information and Communications Technology (ICT) Risk Outcomes," focuses on providing the appropriate attention within an organization's enterprise risk management (ERM) program. This guide was created as a companion to the NIST SP 800-221, "Enterprise Impact of Information and Communications Technology

Risk" that focuses on using risk registers to communicate and manage enterprise risk.

Together, these NIST Special Publications offer a comprehensive suite of guidelines that support organizations in establishing robust cybersecurity programs. They facilitate a standardized approach to cybersecurity, enabling organizations to assess their risks, implement appropriate security controls, and continuously monitor their effectiveness. By adhering to these guidelines, organizations can not only protect their information systems and data, but also contribute to the broader goal of enhancing national cybersecurity resilience.

CENTER FOR INTERNET SECURITY CRITICAL SECURITY CONTROLS (CIS CSC)

The Center for Internet Security (CIS) Critical Security Controls (CSC) is a set of best practices and guidelines designed to help organizations enhance their cybersecurity defenses and reduce the risk of cyber threats. The CIS CSC provides a prioritized and actionable security controls framework that organizations can implement to improve their overall security posture.

The CIS CSC consists of 18 specific controls that cover a wide range of cybersecurity areas, including asset management, secure configuration, continuous vulnerability management, and incident response. These controls are based on real-world attack patterns and are regularly updated to address emerging threats and vulnerabilities. The controls are organized into three implementation groups: Basic, Foundational, and Organizational, which represent progressive levels of security maturity and coverage.

The CIS CSC is a valuable resource for organizations looking to establish a solid foundation of cybersecurity controls. It provides practical and effective measures that can be implemented to mitigate common security risks and enhance the organization's ability to detect, respond to, and recover from cyber incidents. The controls are designed to be adaptable to various environments. They can be tailored to meet different organizations' specific needs and requirements.

CONTROL OBJECTIVES FOR INFORMATION AND RELATED TECHNOLOGY (COBIT 5)

Control Objectives for Information and Related Technology (COBIT 5) is a comprehensive framework that provides guidance and best practices for the governance and management of enterprise information technology systems. COBIT 5 is developed by the Information Systems Audit and Control Association (ISACA) and focuses on aligning information technology activities with business objectives, ensuring effective risk management, and optimizing IT resources.

COBIT 5 defines a set of control objectives that cover various information technology domains, including governance, risk management, strategic alignment, value delivery, and performance measurement. These control objectives serve as targets or desired outcomes that organizations aim to achieve through the implementation of specific controls and processes. They provide a systematic approach to managing information technology risks, improving information technology processes, and ensuring the delivery of value to the organization.

By implementing COBIT 5, organizations can establish a clear governance and management framework for their information technology activities. The framework helps organizations identify and prioritize information technology risks, define control objectives, and implement controls to mitigate those risks. It also provides a structured approach for measuring and monitoring the performance of information technology processes, ensuring that they align with the organization's strategic goals and deliver value.

CYBER RISK INSTITUTE (CRI) PROFILE

The **Cyber Risk Institute (CRI) Profile** was created by the Cyber Risk Institute, a not-for-profit coalition of financial institutions and trade associations whose aim is to protect the global economy through cyber resiliency and standardization. The coalition originally created the CRI profile based on the NIST Cybersecurity Framework version 1.1 and has since fully updated the profile for alignment with version 2.0 of the framework.

The CRI profile has become a benchmark for cybersecurity and resilience in the financial services industry. The profile was created as a concise list of assessment questions that include all of the categories and subcategories from the NIST Cybersecurity Framework as well as other relevant frameworks and standards.

The profile has been designed as an efficient approach to cybersecurity risk management. The CRI profile contains 318 Diagnostic Statements that an organization can use to determine their cybersecurity risk posture and overall cyber resiliency.

INFORMATION TECHNOLOGY INFRASTRUCTURE LIBRARY (ITIL)

ITIL, formerly known as the Information Technology Infrastructure Library, is a widely adopted framework that provides best practices for information technology service management (ITSM). In its latest version, ITIL 4, a holistic approach to service management that focuses on value co-creation, continual improvement, and integrating information technology services with business processes is utilized.

At its core, ITIL focuses on delivering value to customers through the service value system (SVS) to effectively and efficiently manage information technology services to co-create value between a service provider and their customer. It emphasizes the importance of understanding and meeting customer requirements, establishing clear service strategies, designing robust service architectures, and continuously monitoring and improving service delivery.

ITIL defines a range of control objectives and processes covering various aspects of information technology service management through its 34 individual practices, including incident management; problem management; measurement and reporting; change enablement; service level management; and service desk operations. These control objectives outline the desired outcomes and objectives that organizations aim to achieve in each of its 34 practice guides.

By implementing ITIL, organizations can establish standardized service management practices, improve IT operations' efficiency and effectiveness, and enhance customer satisfaction.

PAYMENT CARD INDUSTRY DATA SECURITY STANDARD (PCI-DSS)

The Payment Card Industry Data Security Standard (PCI-DSS) is a security framework developed by major payment card brands to protect cardholder data and ensure the secure handling of payment transactions. It applies to organizations that store, process, or transmit cardholder data, including merchants, service providers, financial institutions, and other entities involved in the payment card ecosystem.

The PCI-DSS provides a set of comprehensive requirements and controls that organizations must adhere to in order to maintain a secure environment for cardholder data. These requirements cover various security aspects, including network security, system hardening, access control, data encryption, vulnerability management, and ongoing monitoring and testing.

The primary goal of the PCI-DSS is to protect cardholder data from unauthorized access, fraud, and misuse. It helps organizations establish a secure infrastructure, implement robust security measures, and maintain a proactive approach to managing security risks associated with payment card transactions.

The PCI-DSS includes specific control objectives and requirements that organizations must meet to demonstrate compliance. These requirements include maintaining secure network configurations, implementing strong access controls, regularly monitoring and testing systems, and maintaining an information security policy.

HEALTH INSURANCE PORTABILITY AND ACCOUNTABILITY ACT (HIPAA)

The Health Insurance Portability and Accountability Act (HIPAA) is a regulatory framework established in the United States to protect individuals' health information's privacy, security, and integrity.

HIPAA sets forth comprehensive standards and requirements for covered entities, such as healthcare providers, health plans, and healthcare clearinghouses, as well as their business associates, to safeguard patient data. HIPAA comprises two key rules: the Privacy Rule and the Security Rule.

The Privacy Rule establishes the rights of individuals regarding their health information and outlines the responsibilities of covered entities in ensuring its confidentiality. It governs the use, disclosure, and access to protected health information (PHI). It grants individuals control over their personal health data.

The Security Rule, on the other hand, focuses on the technical and administrative safeguards that covered entities must implement to protect electronic PHI (ePHI) from unauthorized access, use, or disclosure. It requires the implementation of measures such as access controls, encryption, audit trails, and contingency plans to ensure the confidentiality, integrity, and availability of ePHI.

Complying with HIPAA is crucial for healthcare organizations to safeguard patient privacy, maintain trust, and avoid potential legal and financial consequences. Organizations can establish a comprehensive approach to protecting health information by aligning the NIST Cybersecurity Framework with HIPAA requirements. This integration enables healthcare entities to address cybersecurity risks while ensuring compliance with HIPAA regulations, creating a secure environment for handling sensitive patient data.

NORTH AMERICAN ELECTRIC RELIABILITY CORPORATION (NERC) CRITICAL INFRASTRUCTURE PROTECTION (CIP) STANDARDS

The North American Electric Reliability Corporation (NERC) Critical Infrastructure Protection (CIP) standards are a set of mandatory cybersecurity regulations developed to ensure the reliability and security of the electric grid in North America. These standards are designed to protect critical infrastructure assets and systems within the electric power industry from cyber threats and potential disruptions.

NERC CIP standards comprise a comprehensive framework encompassing a wide range of cybersecurity requirements and controls specifically tailored to the electric utility sector. The standards address various aspects of cybersecurity, including security management, access control, incident response, physical security, and personnel training.

The primary objective of the NERC CIP standards is to establish a consistent and effective cybersecurity posture across the electric power industry. Compliance with these standards is mandatory for entities responsible for operating the bulk electric system, including generation facilities, transmission operators, and distribution utilities.

By integrating the NIST Cybersecurity Framework with the NERC CIP standards, organizations in the electric power sector can enhance their cybersecurity practices and align them with industry-specific requirements. This integration allows utilities to adopt a risk-based approach to identify, assess, and manage cybersecurity risks while ensuring compliance with the NERC CIP standards. The combined use of these frameworks enables utilities to enhance the resilience and reliability of the electric grid, protecting it from cyber threats and maintaining the secure and continuous delivery of electricity to consumers.

FEDERAL RISK AND AUTHORIZATION MANAGEMENT PROGRAM (FedRAMP)

The Federal Risk and Authorization Management Program (FedRAMP) is a government-wide program established to provide a standardized approach for assessing and authorizing federal agencies' cloud computing services and products. FedRAMP aims to ensure the security, privacy, and reliability of cloud services deployed within the federal government by establishing a rigorous risk management and compliance framework.

FedRAMP defines a set of security controls and requirements that cloud service providers must adhere to in order to meet the program's standards. These controls cover various aspects of cloud security, including data protection, access controls, incident response, and continuous monitoring. By implementing these controls, cloud service providers can demonstrate their commitment to safeguarding sensitive government data and infrastructure.

The primary goal of FedRAMP is to streamline the process of assessing and authorizing cloud services, reducing duplication of efforts, and providing a consistent and efficient approach for federal agencies to evaluate and adopt cloud solutions. By leveraging the FedRAMP framework, government agencies can assess the security posture of cloud service providers and make informed decisions about which services meet their specific security and compliance requirements.

Integrating the NIST Cybersecurity Framework (CSF) with FedRAMP allows federal agencies to align their cloud security strategies with industry best practices and standards. The CSF provides a comprehensive set of guidelines and controls that complement the FedRAMP requirements, enabling agencies to develop robust cybersecurity programs and effectively manage risks associated with cloud services. By combining these frameworks, federal agencies can leverage the benefits of cloud computing while ensuring the confidentiality, integrity, and availability of their sensitive information and systems.

CLOUD SECURITY ALLIANCE (CSA) SECURITY, TRUST, ASSURANCE, AND RISK (STAR) REGISTRY

The Cloud Security Alliance (CSA) Security, Trust, Assurance, and Risk (STAR) Registry program is designed to promote transparency and trust in cloud service providers. STAR provides a framework for cloud service providers to self-assess their security practices and disclose relevant information to customers and stakeholders. It enables customers to make informed decisions about cloud services based on the provider's security controls, compliance with industry standards, and overall trustworthiness.

Under the STAR program, cloud service providers can complete a self-assessment questionnaire covering various cloud security domains, including data protection, access management, incident response, and compliance. The questionnaire provides a standardized set of criteria and best practices for assessing and benchmarking the security capabilities of cloud providers.

The STAR program offers different levels of assurance: the Consensus Assessments Initiative Questionnaire (CAIQ) allows providers to conduct a self-assessment and document their security practices, while the

Cloud Security Alliance Security, Trust, Assurance, and Risk (STAR) Registry certification provides independent third-party assessment and certification of a provider's security controls.

By participating in the CSA STAR program, cloud service providers demonstrate their commitment to transparency and accountability in delivering secure cloud services. Customers can refer to the CSA STAR registry to access participating providers' self-assessment reports and certifications, helping them evaluate the security posture of potential cloud service partners.

Integrating the CSA STAR program with the NIST Cybersecurity Framework enhances the assurance and trustworthiness of cloud services. The NIST framework provides a comprehensive set of security controls and guidelines, while CSA STAR offers a mechanism for providers to demonstrate their compliance and adherence to industry best practices. By aligning these frameworks, organizations can assess and select cloud services that meet their security requirements and ensure a higher level of confidence in the security and trustworthiness of their cloud environment.

OPEN WEB APPLICATION SECURITY PROJECT (OWASP)

The Open Web Application Security Project (OWASP) is a nonprofit organization dedicated to improving the security of web applications. OWASP provides a wide range of resources, tools, and best practices to help organizations identify and address vulnerabilities in their web applications.

OWASP focuses on raising awareness about web application security risks and promoting the adoption of secure development practices. They offer guidance on secure coding, vulnerability testing, and secure deployment strategies. One of the key initiatives of OWASP is the OWASP Top Ten, a regularly updated list of the most critical web application security risks. This list serves as a guide for developers and organizations to prioritize their efforts in mitigating common vulnerabilities.

In addition to the OWASP Top Ten, OWASP provides a wealth of resources, including documentation, cheat sheets, code samples, and security testing tools. These resources help developers and security professionals understand and address the various aspects of web application security, such as input validation, authentication, session management, and secure communication.

The integration of OWASP resources and recommendations with the NIST Cybersecurity Framework (CSF) can greatly enhance an organization's web application security posture. By leveraging the best practices and tools OWASP provides, organizations can effectively manage the risks associated with their web applications and protect sensitive data from common security vulnerabilities. The CSF provides a broader framework for managing cybersecurity risks, and by incorporating OWASP's specific guidance, organizations can strengthen their overall security strategy, particularly in the context of web applications.

SUMMARY

The NIST Cybersecurity Framework was developed to support a wide variety of organizations across multiple sectors and industries. As such, it had to remain generic and broad in nature instead of setting up specific and prescriptive controls.

While some people view this as a potential drawback, this is actually one of the biggest benefits of the NIST Cybersecurity Framework (CSF). Since CSF is non-prescriptive, it allows all organizations the flexibility needed to achieve the desired outcomes in whichever way makes the most sense based on their unique business needs and operating environment.

CHAPTER EIGHT

IMPLEMENTATION TIERS

When implementing the NIST Cybersecurity Framework (CSF), there is no strict order in which the core, tiers, and profiles must be implemented and utilized. However, following a logical sequence that ensures a comprehensive and effective implementation is generally recommended.

As a cyber resilience professional, you will determine the most efficient and effective method to implement the framework based on your specific organizational needs and workflows. Many people prefer to choose an implementation tier first, then create a profile based on the selected tier level and utilize the core to determine which categories and subcategories should become a part of your organization's planned implementation, monitoring, and maintenance requirements.

Before you begin to implement the framework using this sequence, it is important that you have a good understanding of the basics of the framework core. Remember, the framework core serves as the foundation for implementing the NIST cybersecurity framework because it outlines the six functions (Govern, Identify, Protect, Detect, Respond, and Recover) and their associated categories and subcategories. If you don't have a solid understanding of the core functions, you will have a hard time adopting the framework and be unable to create a holistic approach to managing cybersecurity risks using the framework.

By following this order of implementation, organizations can progressively build a strong cybersecurity foundation. Remember that the core provides the fundamental functions and activities; the tiers enable the evaluation of the organization's progress in implementing cybersecurity capabilities and practices; and the profiles customize the framework to align with organizational goals and risk management strategies. This implementation is not solely linear but is also an iterative process where organizations may revisit and refine their implementation over time to adapt to evolving threats and changing business requirements.

CHOOSING AN IMPLEMENTATION TIER

Once the core functions are well understood, organizations can assess their cybersecurity capabilities and progress using the framework implementation tiers. These implementation tiers provide a benchmark for evaluating the effectiveness of an organization's cybersecurity practices.

By first assessing the organization's current tier level, you can identify its strengths and areas for improvement to allow for a targeted and strategic approach to enhancing its cybersecurity capabilities.

Choosing an implementation tier step is crucial during the implementation of the NIST Cybersecurity Framework. This step allows organizations to assess their current cybersecurity capability and determine the desired level of cybersecurity practices they aim to achieve. This process involves evaluating the effectiveness of the organization's existing cybersecurity practices and identifying areas that require improvement.

During the assessment, organizations can identify their strengths and weaknesses in various cybersecurity domains, such as risk management, threat intelligence, incident response, and security controls. By understanding their current tier level, organizations gain valuable insights into their cybersecurity posture and can prioritize their efforts accordingly.

When choosing an implementation tier, organizations should consider their risk tolerance, available resources, business objectives, and the cybersecurity capabilities and progress level they aspire to attain. Selecting an implementation tier that aligns with the organization's risk appetite and long-term cybersecurity goals is important. This ensures that the chosen tier represents a realistic and achievable target for the organization.

Recall that the NIST Cybersecurity Framework has four different implementation tiers in which your organization can fall into and be classified. These tiers go from Tier 1 to Tier 4, going from least to most capability in terms of their organizational cybersecurity program.

Tier 1 organizations, also referred to as Partial, have ineffective and inconsistent cyber risk management methods. These organizations have limited awareness of cybersecurity risks and lack systematic risk management processes. Instead, they use an ad hoc and reactive approach to cybersecurity by implementing fragmented practices without a structured approach.

Tier 2 organizations, also referred to as Risk-Informed, have a higher level of cybersecurity implementation. These organizations have developed some formalized policies and procedures and demonstrate a greater level of awareness and proactive cybersecurity practices. Cyber resilience professionals often describe these organizations as having risk management methods that are informal and underdeveloped. Risk informed organizations have an awareness of cybersecurity risks at an organization level, but they lack an organization-wide approach to managing their cybersecurity risks. Their organizational risk management processes are still evolving, and there is still much room for improvement.

Tier 3 organizations, also referred to as Repeatable, have structured risk management methods and well-defined processes in place. These organizations have robust risk management programs and routinely review their risk management participation. Tier 3 organizations demonstrate a proactive approach to cybersecurity, focusing on continuous improvement and effective response to emerging threats. Repeatable organizations have consistent methods in place, which are applied organization-wide to properly manage their cybersecurity risks. Cyber resilience professionals believe these organizations can deliver consistent results from their cyber risk management practices.

Tier 4 organizations, also referred to as Adaptive, have a proactive, innovative, and adaptive approach to cybersecurity. These organizations not only demonstrate the characteristics of Tier 3, but also have an advanced capability to adapt and respond to evolving cybersecurity risks. Tier 4 organizations actively seek out emerging technologies, collaborate with industry partners, and continuously strive for excellence in their cybersecurity posture. Cyber resilience professionals identify these organizations as having risk management methods with feedback loops that aid the organization in learning from experience and organizations that are continually getting better over time. In the NIST Cybersecurity Framework, Tier 4 is considered the highest of the implementation tiers.

Many perceive these tiers as a representation of the organization's progression upward in cybersecurity capabilities and signify their level of effectiveness in managing cybersecurity risks. Organizations can assess their current tier level and set goals to advance to the next tier, gradually improving their cybersecurity practices and enhancing their overall resilience.

IMPLEMENTATION TIER PROGRESSION

Implementation tier progression is critical to enabling organizations to advance their cybersecurity capabilities and progress over time. As organizations assess their current tier level and identify areas for improvement, the implementation tier progression provides a roadmap for strategic growth and enhancement of cybersecurity practices. By understanding the key factors that drive progression from one tier to the next, organizations can establish clear objectives, allocate resources effectively, and develop targeted action plans to elevate their cybersecurity posture.

For example, if an organization is currently at Tier 1 (Partial) with limited awareness and ad hoc cybersecurity practices, they may choose to set their target tier as Tier 2 (Risk Informed). This would involve developing formalized policies and procedures, enhancing awareness and proactive cybersecurity practices, and building a stronger foundation for risk management.

Choosing the right implementation tier sets the stage for organizations to plan and implement the necessary improvements in their cybersecurity practices. It provides a roadmap for enhancing their cybersecurity capabilities and progressing toward a more robust and resilient security posture. The implementation tier acts as a guide to help organizations allocate resources, prioritize initiatives, and track progress as they work toward their cybersecurity goals.

There are three dimensions of cyber risk management that are measured within each tier level, including the risk management process, the integrated risk management program, and external participation.

The dimension of the risk management process within the implementation tier progression refers to the organization's approach and effectiveness in identifying, assessing, mitigating, and managing cybersecurity risks. It encompasses the methodologies, procedures, and practices the organization employs to systematically address risks to its systems, assets, and data. This evaluation helps determine if the organization has a structured and well-defined process in place and if it consistently applies risk management principles to identify, analyze, and respond to cybersecurity risks. During the evaluation of this dimension, the question being addressed is: "How well does the organization establish and execute a risk management process that aligns with its risk tolerance, objectives, and overall business strategy?" and "How well does my organization practice risk management?"

The dimension of the integrated risk management program focuses on the organization's ability to integrate cybersecurity risk management into its overall business processes and decision-making. It assesses the extent to which cybersecurity risk considerations are embedded within the organization's governance structure, policies, and practices. This dimension evaluates whether the organization has established a comprehensive and cohesive program that aligns cybersecurity objectives with its broader strategic goals. By evaluating this dimension, organizations can identify gaps and opportunities for integrating cybersecurity risk management more effectively across their operations, thereby enhancing their overall risk posture. The questions being addressed during the evaluation of this dimension are: "To what extent is cybersecurity risk management integrated into the organization's overall governance and business processes?" and "How repeatable are the outcomes that the organization produces?"

The dimension of external participation focuses on the organization's engagement with external stakeholders, industry collaborations, and information-sharing efforts. It assesses the organization's involvement in relevant cybersecurity communities, its partnerships with other entities, and its participation in information-sharing initiatives. This dimension recognizes the importance of collaboration and the exchange of best practices and threat intelligence in enhancing an organization's cybersecurity capabilities. By evaluating this dimension, organizations can determine the level of their external engagement and identify opportunities for strengthening collaboration with other stakeholders to bolster their cybersecurity defenses. The evaluation question for this dimension is: "To what extent does the organization engage in external partnerships and information sharing to enhance its cybersecurity posture?" and "How well does my organization adapt to new risks and threats?"

MATURITY MODELS

Determining the appropriate implementation tier to target is a crucial step in the implementation of the NIST Cybersecurity Framework (CSF). While the CSF implementation tiers provide a benchmark for evaluating an organization's cybersecurity practices, it is important to note that they are not a traditional maturity model like some other frameworks or models in the industry.

A **maturity model** is a structured framework that assesses and guides the progression of an organization's capabilities and maturity levels in a specific domain, providing a roadmap for improvement and growth. Maturity models typically provide a structured progression of maturity levels, whereas the CSF implementation tiers focus on evaluating the effectiveness of cybersecurity practices rather than the organization's maturity level.

To determine the desired implementation tier, organizations can consider using established maturity models that align with their cybersecurity goals and objectives. These maturity models provide a roadmap for organizations to assess their current state, define target maturity levels, and identify the steps required to progress toward those levels. Four commonly used maturity models are the Capability Maturity Model Integration (CMMI), the ISO/IEC 27001 maturity model, the Cybersecurity Maturity Model Certification (CMMC), and the Cybersecurity Capability Maturity Model (C2M2).

The **Capability Maturity Model Integration** (CMMI) is a widely recognized model that assesses the maturity of an organization's processes across various domains, including cybersecurity. CMMI consists of five maturity levels, moving through from initial, managed, defined, quantitatively managed, and optimizing, and it provides organizations with a framework to assess and improve their process maturity.

The **ISO/IEC 27001 maturity model** is a framework that assesses the maturity level of an organization's information security management system (ISMS) based on the ISO/IEC 27001 standard. This maturity model provides a structured approach for organizations to evaluate their current state of information security practices and measure their progress toward achieving higher levels of maturity. It defines a set of criteria and indicators to assess the effectiveness, efficiency, and sustainability of the organization's ISMS. Using the ISO/IEC 27001 maturity model, organizations can identify areas for improvement, prioritize their efforts, and establish a roadmap for advancing their information security capabilities. The model enables organizations to systematically enhance their security controls, risk management processes, and overall information security posture in alignment with international best practices.

The **Cybersecurity Maturity Model Certification (CMMC)** is a maturity model developed by the U.S. Department of Defense (DoD) to assess and certify the cybersecurity maturity of organizations participating in DoD contracts. The latest version of CMMC only consists of three levels, moving from level 1 (foundational cyber hygiene) to level 2 (advanced cyber hygiene) to level 3 (expert cyber hygiene). Each level represents a different level of cybersecurity maturity and serves as a framework for organizations to align their cybersecurity practices with DoD requirements.

The **Cybersecurity Capability Maturity Model (C2M2)** is a maturity model developed by the U.S. Department of Defense (DoD) to assess and improve an organization's cybersecurity capabilities and maturity. It offers a comprehensive framework encompassing various cybersecurity domains, such as risk management, incident response, secure configuration management, and security awareness. The latest version of the C2M2 evaluates organizations on a three-level scale, ranging from Initiated to Performed to Managed, to determine their level of maturity in cybersecurity practices. Each level represents an increasing level of sophistication and effectiveness in cybersecurity measures. By leveraging the C2M2, organizations can gain valuable insights into their current cybersecurity capabilities, identify areas of weakness for improvement, and establish a clear roadmap for enhancing their cybersecurity posture. The model provides a standardized approach for benchmarking and measuring progress, enabling organizations to prioritize resource allocation, and effectively address critical areas of cybersecurity enhancement.

These maturity models, among others, can complement the NIST CSF implementation tiers by providing organizations with a more detailed and structured approach to assessing and improving their cybersecurity maturity.

It is important to note that the NIST Cybersecurity Framework's implementation tiers are not considered a maturity model. The official guidance from NIST makes this extremely clear, "While organizations identified as Tier 1 (Partial) are encouraged to consider moving to Tier 2 or greater, Tiers do not represent maturity levels."

Instead, while the CSF implementation tiers focus on evaluating the effectiveness of cybersecurity practices, maturity models offer a broader perspective on the overall maturity and capability of an organization's cybersecurity program. By leveraging both the NIST CSF implementation tiers and relevant maturity models, organizations can develop a comprehensive approach to enhance their cybersecurity practices and maturity, aligning with their specific goals and industry requirements.

STRATEGIES FOR MOVING BETWEEN IMPLEMENTATION TIERS

The implementation tiers are the most helpful for setting the tone of the cybersecurity practices within an organization from top executives. When the senior leaders of an organization set a target tier for the organization, such as Tier 2 or Tier 3, this dictates the level of effort and resources that the organization will use toward its cybersecurity and risk management programs.

As previously stated, though, the implementation tiers are not considered a maturity model, so there is no requirement that states an organization must aim to one day become a Tier 4 organization. Instead, your target tier should typically be determined by the organization's unique characteristics and risk tolerance. It is enough to have the organizational leadership state, "Let's be Tier 2". The organization is not required, or even encouraged, to necessarily aim at reaching Tier 3 or 4 unless it makes sense for their organization's needs and risk profile.

As a cyber resilience professional, you may be asked to help an organization move from one tier to another. The best way to approach this is with a five-step process.

First, the organization must assess the current state of its cybersecurity and risk management programs and practices to determine its current implementation tier level. This allows everyone to identify that starting point for the organization once the implementation plan is created.

Second, the organization must define the target state based on the tier level they want to move toward. After all, if the organization doesn't have a target tier or goal selected, the organization can never determine if it has succeeded in its efforts to move between different tier levels.

Third, the organization must develop a plan of action. This plan may include purchasing and installing new technologies, writing new policies, or training existing employees on the organization's current programs and processes. This action plan will depend on the exact situation and challenges the organization is trying to address by setting its target tier level differently than its current tier level.

Fourth, the organization will implement the plan of action. For example, consider an organization that identified itself as a Tier 1 organization and set a target of becoming a Tier 2 organization. In this case, the organization currently has ineffective and inconsistent cyber risk management methods and fragmented practices being deployed using an unstructured methodology. To reach Tier 2, the organization will need to develop formalized policies and create a basic risk management program, even if it is an informal and underdeveloped program compared to a Tier 3 organization.

Fifth, the organization will monitor and adjust. Improving an organization's cybersecurity and risk management programs will not happen overnight, and it is rare to see an organization try to move from Tier 1 to Tier 4 quickly. Instead, this becomes a continual improvement process until the organization reaches the level identified as its target tier. So, if the organization is trying to move from Tier 1 to Tier 2, it will develop a plan, implement a plan, and then monitor to see if the desired results are achieved. If they are not achieving the desired results, then the organization will need to adjust by returning to step three (develop a plan of action) and selecting a new approach to continually improve itself until it reaches the desired target tier initially selected by the organization.

Unfortunately, selecting your target implementation tier is not useful for determining gaps that can be closed within a reasonable amount of time and at a reasonable cost. Too many things could be covered between each tier level, so the available detail received by selecting Tier 2 or Tier 3 as the target tier is simply not actionable enough. To really identify and close these gaps between the tiers, an organization will need to also use profiles built out of the categories and subcategories from the NIST CSF core. Additionally, as a practitioner you will create and implement a Cyber Risk Management Action Plan (CR-MAP) to close the identified gaps using a repeatable and consistent process over time.

SUMMARY

In this chapter, we explored the concept of implementation tiers within the NIST Cybersecurity Framework (CSF) and their role in evaluating an organization's cybersecurity practices. We began by understanding that there is no strict order in which the core, tiers, and profiles must be implemented, but following a logical sequence can lead to a comprehensive and effective implementation.

Remember, there are four implementation tiers: Tier 1 (Partial), Tier 2 (Risk Informed), Tier 3 (Repeatable), and Tier 4 (Adaptive). Each tier represents a level of capability in an organization's cybersecurity practices, with Tier 4 being the highest level of capability. Even though the term maturity is often used heavily by practitioners in the industry when describing the implementation tiers, it is important to note that the implementation tiers are not a maturity model, and it is more accurate to state that the implementation tiers are more focused on capability levels than maturity levels. Instead, organizations can select a compatible maturity model if they wish to certify their maturity, such as the Capability Maturity Model Integration (CMMI), the ISO/IEC 27001 maturity model, the Cybersecurity Maturity Model Certification (CMMC), and the Cybersecurity Capability Maturity Model (C2M2).

Furthermore, we delved into the importance of implementation tier progression and how organizations can strategically advance their cybersecurity capabilities. We emphasized that progression from one tier to the next involves setting clear objectives, allocating resources effectively, developing targeted action plans, and monitoring the results of the plan's implementation. By understanding the dimensions of the risk management process, integrated risk management program, and external participation, organizations can evaluate their current tier level and chart a path toward higher capability. Also, recall that the process of moving between different implementation tiers is not meant to be seen as a linear progression, but instead, it is more agile and spiral in nature as the organization identifies its current practices, attempts to improve them, monitors the results, and then adapts a new plan to continue their improvement in the identified areas.

The implementation tiers in the NIST CSF provide organizations with a framework to assess and enhance their cybersecurity practices. The tiers act as guideposts, allowing organizations to identify their current capabilities, set targets for improvement, and allocate resources effectively. Implementing the framework in a systematic and progressive manner helps organizations build a strong cybersecurity foundation and adapt to evolving threats, ultimately enhancing their overall cybersecurity posture and resilience.

CHAPTER NINE

PROFILES

Imagine you're a master chef preparing a delicious meal. As you gather your ingredients, you realize that each person who will be dining tonight has their own unique tastes and preferences. One person loves spicy flavors, while another prefers mild and savory ones. To satisfy everyone's palate, you customize each dish, adjusting the seasonings, ingredients, and cooking techniques used to prepare the meal.

In the world of cybersecurity, organizations face a similar challenge when implementing the NIST Cybersecurity Framework. While the core functions and implementation tiers provide a solid foundation for cyber resilience, the customization of the framework using profiles is key to aligning the framework with an organization's unique needs, goals, and risk management strategies.

In this chapter, we will explore the concept of profiles within the NIST Cybersecurity Framework. A profile consists of an organization's cybersecurity objectives, current state, and target state, providing a roadmap for aligning cybersecurity activities and priorities with the organization's business requirements.

These profiles allow you to tailor the framework to an organization's specific requirements. Just as a chef may customize each dish, a profile enables you to adjust and fine-tune the framework to effectively address an organization's unique cybersecurity challenges. Essentially, it is a specific and tailored version of the framework core's functions, categories, and subcategories selected just for that organization.

Before diving into the intricacies of profiles, it is essential to have a solid understanding of the framework core and the organization's selected target implementation tier. As we mentioned in the previous chapter, the core functions form the foundation of the NIST Cybersecurity Framework (CSF) by encompassing the six key areas of Govern, Identify, Protect, Detect, Respond, and Recover. These functions and their associated categories provide the building blocks for managing cybersecurity risks. By familiarizing yourself with the core, you can better appreciate how profiles enhance and complement the framework, allowing for a more tailored and targeted approach to cybersecurity.

Throughout this chapter, we will explore the role of profiles in customizing the NIST CSF. We will delve into the process of creating a profile, examining how it aligns with organizational goals and risk management strategies. Additionally, we will discuss the benefits of profiles in promoting effective communication, collaboration, and decision-making within the organization. By the end of this chapter, you will understand how profiles empower your organization to navigate the complex landscape of cybersecurity with a customized approach that enhances your overall cybersecurity posture.

KEY COMPONENTS OF A PROFILE

A profile within the NIST Cybersecurity Framework (CSF) is a customizable tool that allows organizations to align the framework with their specific risk management objectives, industry-specific regulations, and internal priorities. Profiles enable organizations to tailor the framework to their unique needs, ensuring their cybersecurity efforts are focused and effective.

To create a comprehensive and meaningful profile, it is important to consider a profile's three key components: core functions, categories, and subcategories.

First, the core functions. The core functions are the foundation of the framework and provide a structured approach to managing cybersecurity risks. The six functions of Govern, Identify, Protect, Detect, Respond, and Recover encompass a wide range of activities and outcomes that organizations need to address in their cybersecurity practices. When creating a profile, evaluating each core function and determining its relevance and priority to a specific organization is essential. By understanding and selecting the core functions that align with your risk management objectives, you can effectively tailor the profile to meet the organization's unique needs.

Second, the categories. Within each core function, the framework defines various categories that further break down into subcategories. These categories provide a more granular view of the specific areas organizations should consider when managing their cybersecurity risks. Categories include asset management, risk assessment, awareness and training, data security, and many others. When developing a profile, it is important to identify the relevant categories that align with your organization's priorities and risk management strategies. By focusing on these specific categories, you can customize your profile to address the areas that are most critical to your organization's cybersecurity posture.

Third, the subcategories. Under each category, defined subcategories represent specific outcomes within each category. These subcategories provide further detail and guidance on which outcomes organizations should undertake to achieve desired cybersecurity objectives. When developing a profile, reviewing the subcategories within the selected categories and assessing their relevance and applicability to your organization's context is important. By identifying the specific subcategories that align with your risk management goals, you can customize your profile to include the activities and outcomes that are most relevant to your organization's cybersecurity practices.

CREATING A PROFILE

In cyber resilience, a one-size-fits-all approach rarely yields the best results. Each organization around the world faces unique challenges, operates in specific industries, and has distinct risk management objectives. By creating a profile from the framework, organizations can customize the framework to their specific requirements, aligning their cybersecurity efforts with their unique risk landscape and business priorities. This unique profile that the organization creates will be used to establish a roadmap that aligns its cybersecurity practices with its risk management strategy to enhance its resilience against cyber threats and enables effective protection of its critical assets and operations.

To create a profile, the cyber resilience professional should follow a basic six-step process to identify organizational requirements and objectives; evaluate the framework core; select and prioritize categories; define subcategories; establish performance goals and metrics; and document and implement the profile.

The first step is to identify organizational requirements and objectives. You should begin by thoroughly assessing the organization's risk management objectives, industry-specific regulations, business priorities, and cybersecurity needs. It is important to identify the key drivers that shape the organization's cybersecurity strategy, such as protecting sensitive data, complying with regulatory requirements, or safeguarding its critical infrastructure. This step is crucial in understanding the organization's unique cybersecurity requirements and setting the foundation for creating a tailored profile.

The second step is to evaluate the framework core. As stated previously, the cyber resilience professional must be familiar with the framework core and its six functions (Govern, Identify, Protect, Detect, Respond, Recover), categories, and subcategories. You should assess how each core component aligns with the organization's objectives and its risk management strategy. With the help of key stakeholders, you should identify the specific categories and subcategories that are most relevant and critical to the organization's cybersecurity efforts. This evaluation will help determine the core components that must be included in the tailored profile you are now creating.

The third step is to select and prioritize categories. Based on the evaluation conducted in the second step, you should select the categories that align with the organization's priorities and risk management objectives. Considering the unique cybersecurity challenges and focus areas that are specific to the organization, you must prioritize these categories based on their importance and relevance to the organizational cybersecurity strategy. This step ensures that the created profile is tailored to address the most critical aspects of the organization's cybersecurity needs because no organization has enough time or money to focus on everything all of the time. Choices will have to be made, and that is why selecting and prioritizing the categories becomes so important.

The fourth step is to define subcategories. Within each selected category, define the specific subcategories that will guide the organization's cybersecurity efforts. These subcategories provide a more granular level of detail and actionable steps to achieve the desired cybersecurity outcomes. Customize these subcategories to align with the organization's specific needs, resources, and risk landscape. This step ensures that the newly created, tailored profile includes the necessary measures and controls to address the organization's cybersecurity challenges effectively.

The fifth step is to establish performance goals and metrics. Setting performance goals and metrics will allow the organization to measure the effectiveness and progress of its cybersecurity practices. These goals and metrics should align with the organization's risk appetite, compliance requirements, and overall cybersecurity strategy. Also, take the time to define clear indicators that will help to evaluate the success and capability of the organization's cybersecurity program. By establishing performance goals and metrics, you can monitor the organization's cybersecurity posture, track its improvements over time, and make informed decisions to enhance its cybersecurity capabilities.

The sixth step is to document and implement the profile. Documenting the customized profile helps capture the selected categories, subcategories, performance goals, and metrics for later review and analysis. By ensuring that the profile is well-documented, easily understood, and accessible to your organization's relevant stakeholders, you can more easily communicate the profile to the key personnel responsible for implementing and managing the organization's cybersecurity program. During this step, it is also important to actually conduct the implementation of the profile by integrating the defined measures, controls, and practices into the organization's cybersecurity operations. After the initial implementation is complete, it is important to periodically review and update the profile as the organization's cybersecurity needs evolve over time.

By creating a profile from the NIST Cybersecurity Framework, organizations are empowered to tailor their cybersecurity efforts and focus their actions on their specific risk landscape and priorities. By following this step-by-step process, organizations can create a profile that aligns with their unique requirements, enhances their cybersecurity posture, and enables proactive risk management.

PROFILE TAILORING

A framework profile can measure the gap between the current state of cybersecurity practices and the target state, including the target implementation tier selected by the organization. As the organization begins to create its profile, it may discover that removing some subcategories or entire categories from the definition of the organizational target state is needed.

For example, you might be creating a profile for the framework to focus on how well an organization performs the Detect function while making the conscious decision to leave out the other five functions until the organization has more time, resources, and effort available to focus on them. Alternatively, you might focus on only the ten subcategories for the Cybersecurity Supply Chain Risk Management (GV.SC) activity and ignore every other category and subcategory in the framework core for the time being. Either of these are acceptable ways to create a scoped-down target profile for an organization if you intend to conduct a limited engagement with them.

Consider a tailored profile for a certification examination institute:

> **Overall Objective:**
> To enhance the protection of sensitive data and ensure the secure access and authentication of users in the certification exam systems.
>
> **Profile Categories:**
> 1. Protect (PR) function
>   - PR.AA: Identity Management, Authentication, and Access Control
>   - PR.AT: Awareness and Training
>
> **Profile Objectives:**
> - Establish robust identity management, authentication, and access control measures to safeguard sensitive data in our certification exam systems
> - Enhance cybersecurity awareness and provide effective training to our personnel and users to mitigate the risk of unauthorized access and security incidents
>
> **Profile Subcategories:**
> - PR.AA-03: Users, services, and hardware are authenticated to ensure secure access to certification exam systems
> - PR.AT-01: Personnel are provided with awareness and training so that they possess the knowledge and skills to perform general tasks with the cybersecurity risks of the certification exam systems in mind
>
> **Profile Performance Goals:**
> - Increase the usage of multi-factor authentication for privileged user system access
> - Achieve high completion rates for cybersecurity awareness training modules
> - Reduce the number of unauthorized access incidents
>
> **Profile Metrics:**
> - Percentage of privileged users utilizing multi-factor authentication
> - Completion rates of cybersecurity awareness training modules
> - Number of unauthorized access incidents reported

Since we assume an extremely limited engagement with the organization for this example, we have developed a tailored profile by identifying the most important items from the larger NIST Cybersecurity Framework that address this organization's key areas of concern: the Protect function, the PR.AA and PR.AT categories, and the PR.AA-03 and PR.AT-01 subcategories.

As you read through the sample target profile, you should have noticed several interesting things about it.

Notice that when the profile was created, it only used one of the possible six functions from the NIST Cybersecurity Framework. Remember that the framework is voluntary, and the organization can use as much or as little of it as they desire. In this example, the organization and its consultants have only focused on the Protect function in this profile.

In the second section, the organization and its consultant have decided to only adopt a profile with two categories: Identity Management, Authentication, and Access Control (PR.AA) and Awareness and Training (PR.AT). While there may have been other subcategories that could have been added to help meet the overall organizational objective, the decision was made to limit the scope of the profile and the number of subcategories selected to only the most important for this particular engagement. This decision may have been made due to the amount of time and resources available from the organization for its cybersecurity program and risk management programs, or it could have been because the organization was extremely new and just trying to establish its initial actions in a new cybersecurity program.

In the third section, you will notice that the profile objectives were not written exactly as they are presented in the NIST Cybersecurity Framework official documentation. In the target profile, it states that the organization will "Establish robust identity management, authentication, and access control measures to safeguard sensitive data in our certification exam systems." This most closely aligns with the category of PR.AA which states that "access to physical and logical assets is limited to authorized users, services, and hardware, and managed commensurate with the assessed risk of unauthorized access."

The important concept here is that you can tailor your profile however you see fit. In this case, the objectives were tailored to have a very narrow focus and to be written in a more business-like and non-technical tone so that they could be more easily communicated to the organization's executives.

In the fourth section, if you cross-reference the PR.AA-03 and PR.AT-01 subcategories from the official NIST Cybersecurity Framework documentation with the tailored profile provided, you will find that they are not a direct match word for word. This, again, is part of the tailoring.

For example, PR.AA-03 in the official documentation is written as "Users, services, and hardware are authenticated." In the tailored profile, this was written as "Users, services, and hardware are authenticated to ensure secure access to certification exam systems." The difference here is that the tailored version is more specific and even provides a potential solution that the organization seeks to implement.

In the final two sections, you will notice the performance goals and metrics. These are not found inside the NIST Cybersecurity Framework itself, but they are critically important when developing a targeted profile for an organization. The performance goals state what we are trying to achieve with the actions specified in the profile. These tend to be more generic in nature, such as "reduce the number of unauthorized access incidents."

The metrics, on the other hand, indicate how that performance goal will be measured. In this case, the target profile states that this performance goal will be measured by counting the number of unauthorized access incidents reported.

Notice that the target profile did not include the specific number to be used as a goal for this metric. This is because we want to keep this target profile generic enough that we can reuse it over the years. During the kick-off of the organization's action plan, the metrics will be given a specific target number that the organization is working toward. At this point, though, we don't know how many incidents the organization has had over the last 12 months, so it is impossible to provide an exact number for the metric while drafting up the target profile.

Please note this tailored profile is an extremely simplified profile created for illustrative purposes. Organizations should customize their profiles based on their specific needs, risk landscape, and cybersecurity objectives. A typical profile can end up being anywhere from 20 to 100 pages, or more, in length.

PROFILE TEMPLATES

Creating a profile within the NIST Cybersecurity Framework can be a complex and time-consuming task, requiring careful consideration of an organization's unique needs, risk landscape, and cybersecurity objectives.

A profile template can be utilized to simplify the profile creation process and provide organizations with a starting point. These templates serve as pre-defined frameworks that outline a profile's key components and structure, allowing organizations to customize and tailor them according to their specific requirements.

Profile templates provide several benefits, including consistency, efficiency, and alignment with recognized cybersecurity best practices. They offer a structured framework that guides organizations through the profile creation process, ensuring that essential elements are included, and relevant categories and subcategories are considered. Templates also promote consistency across organization or industry profiles, facilitating easier benchmarking, sharing of best practices, and collaboration.

While profile templates can vary based on organizational needs, some common elements are typically included, such as organizational information, profile summary, profile components, profile objectives, profile activities, performance goals, and metrics.

The organizational information should be clearly listed in the profile template. This helps identify the organization's name and relevant details to provide context and ownership.

The profile summary section provides a concise overview of the profile. It helps to summarize the key objectives and focus areas.

The profile components section outlines the specific framework components that the profile will address, such as functions, categories, and subcategories.

The profile objectives section articulates the goals and intentions of the profile, highlighting the desired outcomes and improvements to be achieved.

The profile subcategories section lists the specific activities or actions the organization will undertake to meet the described profile objectives.

The performance goals are used to provide clear and measurable goals, which are established to assess the success and effectiveness of the profile implementation.

The metrics section provides quantifiable metrics to track progress, measure performance, and evaluate the impact of the profile on the organization's cybersecurity practices.

Organizations can leverage existing profile templates available from reputable sources, such as industry associations, government agencies, or cybersecurity frameworks, in addition to creating create their own library of profile templates based on previous engagements at organizations with whom they worked. These templates can serve as a starting point, providing a framework that can be customized and tailored to fit the organization's specific needs and requirements.

Organizations can expedite the profile creation process by utilizing profile templates, ensuring consistency across profiles, and leveraging established best practices. It enables organizations to focus their efforts on customizing the template to their specific context and cybersecurity goals to align with the organization's risk landscape, industry regulations, and internal priorities while still achieving a robust and effective tailored profile.

SECTOR-SPECIFIC PROFILES

In the diverse landscape of cybersecurity, different sectors and industries face unique challenges and have specific requirements for protecting their critical assets and infrastructure. To address these sector-specific needs, the NIST Cybersecurity Framework provides the flexibility to develop tailored profiles that align with specific sectors' cybersecurity objectives and risk landscapes. In addition to generic profile templates, many sector-specific profiles, also known as **community profiles**, have been created, including: CRI Profile for the Financial Sector; Manufacturing Profile; Election Infrastructure Profile; Satellite Networks Profile; Smart Grid Profile; Connected Vehicle Profile; Payroll Profile; Maritime Profile; and Communications Profile.

Each profile focuses on the cybersecurity considerations and controls that are most relevant to the respective sector, providing organizations with guidance and best practices specific to their industry. By leveraging these sector-specific profiles, organizations can enhance their cybersecurity posture and effectively mitigate sector-specific risks, ensuring the resilience and security of critical systems and operations.

CRI PROFILE FOR THE FINANCIAL SECTOR

The Cyber Risk Institute (CRI) Profile was specifically designed by a consortium of financial institutions and trade associations to provide a tailored profile that could address the unique cybersecurity challenges faced by financial institutions. This profile was one of the first to be updated to be in alignment with the NIST Cybersecurity Framework version 2.0 and emphasizes the sector's need for more robust cybersecurity measures to protect against increasingly sophisticated cyber threats against the integrity, confidentiality, and availability of an organization's financial data.

Financial institutions operate in a highly interconnected environment that relies on complex information technology infrastructures to not only manage routine financial transactions, but also to secure their sensitive customer information from cyber attacks. The financial sector's infrastructure is extremely diverse and encompasses everything from traditional banking systems to innovative financial technologies (FinTech). Each of these presents unique vulnerabilities and risk exposures that must be considered by the CRI Profile.

Since financial services play a critical role in both national and global economies, and since financial institutions control the movement of large amounts of money, their systems are considered to be a high-value target for cybercriminals. This requires that cybersecurity professionals utilize a proactive and adaptive approach to their organization's cybersecurity to prevent becoming the next victim of a highly targeted attack against the financial sector's underlying infrastructure.

The CRI Profile integrates and expands upon existing cybersecurity standards, including those set forth by the Financial Services Sector Coordinating Council (FSSCC) and the International Organization for Standardization (ISO), to offer a comprehensive set of cybersecurity controls and best practices tailored to the financial sector's specific needs. This profile addresses many key focus areas, including identity and access management, threat intelligence sharing, fraud detection, and incident response. The profile also places a strong emphasis on risk management and regulatory compliance.

The CRI Profile serves two critical functions: aiding financial institutions in articulating their cybersecurity requirements to third-party service providers and enabling a gap analysis against the NIST Cybersecurity Framework to identify areas for continual improvement.

This profile categorizes cybersecurity outcomes according to the sector's unique risk thresholds by defining three impact levels (low, medium, and high) for each outcome to guide the implementation of appropriate controls. For example, when an organization is managing the risk associated with unauthorized access to financial systems, the profile recommends a tailored set of controls based on the potential impacts to financial stability, customer trust, and regulatory compliance. This approach is more nuanced since it relies on the targeted profile to ensure that financial institutions can prioritize their cybersecurity investments more effectively. This allows the organization to measure the return on their investments while ensuring they achieve higher levels of cyber resilience against cyber threats and potential cyber attacks.

The CRI Profile is a strategic tool for financial institutions that provides a sector-specific roadmap for navigating the complex cyber risk landscape. By emphasizing the importance of collaboration, compliance, and continuous improvement in fostering a secure and resilient financial ecosystem, the CRI Profile can increase the safety and security of the interconnected financial industry.

MANUFACTURING PROFILE

In October 2020, NIST published a Manufacturing Profile based on version 1.1 of the NIST CSF. This document is officially called the "National Institute of Standards and Technology Internal Report 8183 Revision 1". If you want to download a copy, you can find it at https://doi.org/10.6028/NIST.IR.8183r1. Note, at the time of this publication, this is the current version, but NIST is working on updating all of their profiles to become in alignment with NIST Cybersecurity Framework version 2.0.

Manufacturers are a unique industry from a cybersecurity perspective because they have the traditional IT infrastructure to automate routine office tasks, like writing documents and processing emails. But they also use computers to support their manufacturing operations, whether process-based, discrete-based, or some combination of both.

These special-purpose computers are broadly known as industrial control systems (ICS). There are different types of control systems, including supervisory control and data acquisition (SCADA) systems, distributed control systems (DCS), and programmable logic controllers (PLC). If an industrial control system stops working, this also causes the production line to have a work stoppage, too. This means that no work can be done, causing the flow of revenue to stop as well. Manufacturing isn't just used to create large, complex machines like cars, airplanes, televisions, and computers. Manufacturing also includes most things we consume daily, including cans of soda, a plastic bag filled with apples, or even the paper bag where your takeout food is placed.

Because of the heavy emphasis on industrial control systems in manufacturing, the Manufacturing Profile better incorporates ISA/IEC 62443 as the basis for the controls needed to reliably produce the desired outcomes. The ISA/IEC 62443 is an international series of standards on IT security for networks and systems for Industrial communication networks. The International Society of Automation (ISA) and the International Electrotechnical Commission (IEC) are two organizations that worked together to create the ISA/IEC 62443 standard.

There are two particularly good use cases for this specific profile. The first is to help an organization express its cybersecurity risk management requirements to an external service provider. The second is to help the organization compare its current profile to the Manufacturing Profile in order to reveal gaps or weaknesses that might result in its systems being exploited.

The Manufacturing Profile defines three impact levels for each subcategory: low, medium, and high. The profile also defines five impact categories examples within it, including injuries to people, financial loss, environment release, interruption of production, and damage to public image.

Each subcategory presented in the profile has more or fewer controls recommended depending on the impact level. This is easy to understand if you think about the impact of an industrial control system's failure to milk production as compared to refining gasoline: sour milk will get thrown out, but a gasoline spillage could cause an explosion that might kill somebody.

To demonstrate how the Manufacturing Profile could be used, let's consider how one subcategory is profiled within it. It is important to note that at the time of this writing, the Manufacturing Profile still refers to the NIST Cybersecurity Framework version 1.1. The following example will rely upon this nomenclature.

The subcategory is known as ID.AM-1 is defined as "physical devices and systems within the organization are inventoried." This subcategory has been updated in version 2.0 of the framework as "inventories of hardware managed by the organization are maintained." As you can see, these two subcategories essentially say the same thing, so this item from the Manufacturing Profile can also be used within your NIST Cybersecurity Framework version 2.0 implementation.

In the Manufacturing Profile, if a failure to document the inventory of the manufacturing system occurs, this would result in a low impact according to the profile. With a low impact, the following informative references and associated controls are recommended within the Manufacturing Profile:

(1) ISA/IEC 62433-2-1:2009 4.2.3.4

(2) ISA/IEC 62433-3-3:2013 SR 7.8

(3) CM-8

If a medium impact level is expected, then the profile recommends that the organization add CM-8 (1)(3)(5) from the NIST SP800-53 rev.5 as an additional control to further help control the risk. The CM-8 (1) control is focused on the organization updating their inventory of their information system components during installations, updates, and system removals. The CM-8 (3) control is focused on the organization automating the detection of unauthorized components. The CM-8 (5) control is focused on ensuring that there is not duplicate accounting for the organization's information system components within its inventory. Simply put, the organization should be conducting configuration management and maintaining an updated and accurate inventory of their hardware devices.

If a high impact level is determined to exist, then CM-8 (2)(4) from the NIST SP800-53 rev.5 are recommended as an additional control for implementation by the organization. The CM-8 (2) control is focused on employing automated methods of maintaining an up-to-date, complete, and accurate inventory of the organization's information system components. The CM-8 (4) control is focused on ensuring that the organization includes accountability information for each component within their inventory. Simply put, the organization must conduct their configuration management using automated techniques to ensure the inventory remains up-to-date and accurate, and each component should be assigned to a person, position, or role who is responsible for maintaining each information system component.

← ELECTION INFRASTRUCTURE PROFILE

The Election Infrastructure Profile is specifically designed to address the unique cybersecurity challenges faced by organizations involved in electoral processes. This profile aims to ensure the integrity, security, and resilience of the election infrastructure, which includes systems, networks, and data during publicly held elections within a city, state, or country.

The Election Infrastructure Profile focuses on several key areas of cybersecurity to safeguard the election process. It emphasizes the identification and protection of critical assets, such as voter registration databases, voting machines, and communication networks, to prevent unauthorized access or tampering. The profile also emphasizes the importance of continuous monitoring and detection of potential cybersecurity incidents, allowing timely responses to any threats or anomalies. Additionally, the profile encourages robust incident response and recovery capabilities to ensure quick restoration of services and the integrity of the electoral process.

By implementing the Election Infrastructure Profile, election organizations can enhance their cybersecurity posture, strengthen public trust in the electoral system, and safeguard the democratic process. This profile serves as a valuable resource for election stakeholders, providing guidance and best practices to address the unique cybersecurity risks and challenges faced in the context of elections.

SATELLITE NETWORKS PROFILE

The Satellite Networks Profile is designed to address the specific cybersecurity considerations and challenges faced by organizations operating satellite networks. As satellite networks play a crucial role in various industries such as telecommunications, broadcasting, and remote sensing, ensuring the security and resilience of these networks is of the utmost importance.

– The Satellite Networks Profile focuses on key cybersecurity areas to protect the integrity, confidentiality, and availability of satellite systems and their associated data. It emphasizes the identification and protection of critical assets, including ground stations, satellites, and communication links, to mitigate the risk of unauthorized access or interference. The profile also emphasizes the implementation of robust monitoring and detection mechanisms to identify and respond to potential cybersecurity incidents that could impact the functionality and reliability of the satellite network.

By implementing the Satellite Networks Profile, organizations operating satellite networks can enhance their cybersecurity capabilities, safeguard the integrity of their operations, and maintain the trust of their customers and stakeholders. This profile provides valuable guidance and best practices tailored to the unique challenges faced by satellite network operators, enabling them to address cybersecurity risks and maintain the resilience of their systems in an ever-evolving threat landscape.

SMART GRID PROFILE

The Smart Grid Profile addresses the specific cybersecurity concerns and requirements related to the operation and management of smart grid systems. As the energy sector increasingly adopts advanced technologies to improve efficiency and reliability, protecting the smart grid infrastructure becomes crucial for ensuring the continuous and secure delivery of electricity.

The Smart Grid Profile focuses on key areas of cybersecurity to address the unique challenges faced by smart grid systems. It emphasizes the identification and protection of critical assets, including control systems, data centers, communication networks, and IoT devices, to mitigate the risk of unauthorized access or disruption. The profile also highlights the need for robust monitoring and detection capabilities to quickly identify and respond to potential cybersecurity incidents that could impact the reliable functioning of the smart grid.

By implementing the Smart Grid Profile, organizations operating smart grid systems can enhance their cybersecurity posture and resilience, ensuring the secure and reliable delivery of electricity to consumers. The profile provides valuable guidance and best practices tailored to the specific requirements of smart grid systems, enabling organizations to effectively manage cybersecurity risks and maintain the integrity and availability of their critical infrastructure.

CONNECTED VEHICLE PROFILE

The Connected Vehicle Profile addresses the cybersecurity challenges and considerations specific to the connected vehicle ecosystem. With the increasing integration of advanced technologies and connectivity in vehicles, ensuring the security and privacy of connected vehicles becomes paramount for maintaining the safety and trust of passengers and road users.

The Connected Vehicle Profile focuses on safeguarding the critical components and communication networks within connected vehicles. It emphasizes the need for secure authentication and access controls to prevent unauthorized access to vehicle systems and data. The profile also highlights the importance of secure communication protocols and encryption to protect the integrity and confidentiality of data transmitted between vehicles and infrastructure.

By implementing the Connected Vehicle Profile, organizations in the automotive industry can enhance the cybersecurity of their connected vehicle systems. The profile provides specific guidance and controls to address the unique risks associated with connected vehicles, promoting the adoption of best practices for secure vehicle-to-vehicle and vehicle-to-infrastructure communications. By implementing the recommended cybersecurity measures, organizations can ensure the safety, privacy, and resilience of connected vehicles, contributing to the overall advancement and acceptance of connected and autonomous transportation.

PAYROLL PROFILE

The Payroll Profile focuses on the specific cybersecurity challenges and risks associated with payroll systems and processes. Payroll systems play a critical role in organizations as they handle sensitive employee data and financial information. Safeguarding the confidentiality, integrity, and availability of this data is crucial to protect employee privacy and prevent financial fraud or unauthorized access.

The Payroll Profile emphasizes the need for strong access controls and identity management to ensure that only authorized personnel can access payroll systems and data. It also emphasizes the importance of data encryption and secure transmission protocols to protect sensitive information during transit. Additionally, the profile highlights the significance of regular monitoring, detection, and response mechanisms to identify and mitigate any potential payroll system vulnerabilities or unauthorized activities.

By implementing the Payroll Profile, organizations can enhance the security of their payroll systems and protect sensitive employee information. It provides a set of controls and best practices tailored specifically to address the unique cybersecurity risks faced by the organization's payroll functions. By adopting these measures, organizations can ensure the accuracy and confidentiality of payroll data, promote employee trust, and mitigate the potential financial and reputational impacts of payroll-related security incidents.

MARITIME PROFILE

The Maritime Profile addresses the cybersecurity challenges and risks specific to the maritime industry. The maritime sector encompasses a wide range of activities, including shipping, ports, offshore operations, and maritime transportation. As technology advances and digitalization becomes more prevalent in the maritime domain, ensuring the security and resilience of critical maritime infrastructure and systems becomes extremely important for the safety of the industry and the people working within it.

The Maritime Profile focuses on key areas such as vessel and facility security, navigational systems, port operations, and communication networks. It emphasizes the need for robust access controls, authentication mechanisms, and encryption protocols to protect critical maritime assets and data from unauthorized access or tampering. Additionally, the profile underscores the significance of continuous monitoring, incident response planning, and information sharing to detect and respond to potential cyber threats in real time.

By implementing the Maritime Profile, organizations in the maritime sector can strengthen their cybersecurity posture and safeguard their critical assets, operations, and data. This profile provides tailored controls and recommendations that address the unique maritime cybersecurity risks, helping organizations build a resilient cybersecurity framework. With effective cybersecurity measures in place, the maritime industry can ensure the safety and security of vessels, ports, and associated infrastructure, maintain smooth operations, and protect against potential cyber incidents that may disrupt maritime activities or compromise maritime safety.

COMMUNICATIONS PROFILE

The Communications Profile is designed to meet the specific cybersecurity needs and challenges within the communications sector. As the communications industry becomes increasingly reliant on digital technologies and interconnected networks, it is crucial to protect the integrity, confidentiality, and availability of critical communications infrastructure and services.

The Communications Profile focuses on areas such as network security, data protection, threat intelligence, and incident response. It highlights the importance of implementing strong access controls, encryption mechanisms, and secure network architectures to safeguard sensitive information and prevent unauthorized access. Additionally, the profile emphasizes the need for continuous monitoring, vulnerability assessments, and proactive threat detection to identify and mitigate potential cyber risks.

By adopting the Communications Profile, organizations in the communications sector can enhance their cybersecurity resilience and mitigate the potential impacts of cyber threats. It provides industry-specific guidance and recommendations for implementing effective cybersecurity controls, ensuring the reliability and security of communication networks, services, and customer data. With a robust cybersecurity framework in place, the communications industry can foster trust, maintain the confidentiality of sensitive communications, and protect against cyberattacks that may compromise network infrastructure or disrupt communication services.

CURRENT PROFILE VERSUS TARGET PROFILE

The current profile and target profile are essential components of the framework that organizations use to assess their cybersecurity posture and establish a roadmap for improvement.

The current profile represents the organization's existing cybersecurity practices, including its current cybersecurity activities, desired outcomes, and risk management approaches. It provides a snapshot of the organization's current state of cybersecurity, highlighting strengths, weaknesses, and areas for improvement. By understanding the current profile, organizations can identify gaps in their cybersecurity defenses and determine the necessary steps to enhance their security posture.

On the other hand, the target profile represents the organization's desired state of cybersecurity practices and outcomes. It outlines the specific cybersecurity improvements and goals that the organization aims to achieve. The target profile is aligned with the organization's risk management objectives, industry-specific regulations, and internal priorities. By comparing the current profile with the target profile, organizations can identify the gaps and prioritize the actions needed to bridge those gaps and move toward their desired cybersecurity state.

The process of transitioning from the current profile to the target profile involves assessing the organization's current cybersecurity practices, identifying areas for improvement, and developing a strategic action plan. It requires a comprehensive understanding of the organization's risk landscape, business objectives, and available resources. The target profile serves as a guide to help organizations align their cybersecurity practices with their overall business strategy and risk management goals. It provides a roadmap for prioritizing cybersecurity initiatives, allocating resources effectively, and tracking progress toward the desired cybersecurity state.

PROFILES FOR REGULATORY COMPLIANCE

If your organization works in a highly regulated industry, tailoring profiles to support its regulatory compliance requirements is an important aspect of implementing the NIST Cybersecurity Framework. Regulatory compliance mandates specific security controls and practices that organizations must adhere to in order to meet legal and industry-specific obligations. By tailoring profiles to support regulatory compliance requirements, organizations can ensure that their cybersecurity efforts align with the necessary regulatory standards to avoid fines and other negative consequences.

. When tailoring profiles for regulatory compliance, organizations need to carefully assess the specific requirements and guidelines set forth by the relevant regulatory bodies. This involves understanding the regulatory landscape, identifying the applicable regulations, and determining the specific cybersecurity controls and practices mandated by those regulations. Organizations can then align their profiles to include the necessary categories and subcategories that address their regulatory compliance requirements.

Tailoring profiles to support regulatory compliance also involves incorporating industry-specific regulations and standards into the profile. Many industries have their own specific cybersecurity requirements and best practices that organizations must adhere to. By customizing the profile to include these industry-specific regulations and standards, organizations can ensure that their cybersecurity practices not only meet regulatory compliance but also address the unique challenges and risks associated with their specific industry.

Furthermore, tailoring profiles to support regulatory compliance helps organizations establish a structured and systematic approach to meeting their compliance obligations. By mapping the regulatory requirements to the categories and subcategories in the profile, organizations can identify gaps in their current cybersecurity practices and develop a targeted plan to address them. This tailored approach enables organizations to demonstrate their commitment to regulatory compliance, enhance their security posture, and mitigate potential risks associated with non-compliance.

SECURE ONCE AND COMPLY MANY

- Using the framework core will help you focus the organization on doing the activities and outcomes that make the biggest difference toward achieving cyber resilience. There are many things you could do to become more cyber resilient. You have a lot of technical choices, but there are also non-technical possibilities, too.

Most organizations also have a lot of constraints in terms of budget limitations and competing priorities; they simply cannot do everything they might desire to increase their security. So, of all the things you could do to become more resilient, you need to consider which ones will give the organization the biggest return on their investment of resources.

Well, you can use the framework core to take out a lot of the guesswork involved in making that decision. Use of the CSF core can simplify and turbocharge your organization's cybersecurity program by combining it with a "secure once and comply many" approach. To do this, you need to map all of the organization's compliance requirements back to the subcategories in the framework profile.

One place to start identifying the current mandates is within the informative references' column of the framework core, such as those listed as ISO 27001, since this is an international standard on how to manage information security in an organization. Then, add any other compliance mandates, including laws, regulations, and data security addendums, from the organization's contracts.

Anytime new mandates arrive, they should be checked against the current framework profile to see if the organization is already meeting those compliance requirements. If the organization isn't, you may need to add more controls or even a new row into its framework profile to account for everything required of the organization.

Essentially, using this approach, you are creating a compliance architecture. **Compliance architecture** refers to the structure and framework that organizations establish to ensure adherence to regulatory and legal requirements related to cybersecurity and data privacy. This involves designing and implementing policies, processes, controls, and technologies that enable the organization to meet its compliance obligations.

This practice alone will substantially reduce any disruption to your workforce, as you only need to deploy any new controls above and beyond your existing profile's control set.

The main idea behind this 'secure once and comply many' approach is to make it easier to operationalize the framework while including all of the other cybersecurity obligations the organization might have based on their industry or business model. This allows the organization to spend less time and money implementing controls because often a single control can satisfy multiple requirements.

Another benefit of this approach is that the organization will get more consistent execution during the implementation phase since they only need to learn to use one control, regardless of how many requirements that one control satisfies. For instance, the organization may implement a single control that could prevent the need for having one method for deleting PCI-DSS data and a different method for deleting sensitive data regulated by HIPAA.

Even better, the organization's staff can also save lots of time because new data protection requirements are first analyzed centrally to discover any duplication of effort before any changes are rolled out.

Consider the following diagram showing how the different components of a compliance architecture work together under a 'secure once and comply many' approach:

SECURE ONCE AND COMPLY MANY

Notice the center of the circle is surrounded by the NIST Cybersecurity Framework version 2.0. It includes the core and its 22 categories and 106 subcategories.

Around the outside of the circle are all the other cybersecurity and data protection requirements that the organization may have. In this example, this organization is trying to comply with the ISO 27001 requirements, HIPAA regulatory requirements, information security policy, and customer requirements. By mapping these different requirements back to the framework core, duplication can be detected and removed to provide a singular set of controls to meet all four sets of requirements at once.

The innermost circle of the diagram indicates this consolidated set of controls that are created as a tailored framework profile for the organization. This is then used to create the standards, processes, and procedures the organization will utilize to meet the controls selected within its target profile.

Let's work through a real-world example of a company that is affected by HIPAA regulations because they are a HIPAA business associate. A HIPAA business associate is a designation given to a company that isn't in the healthcare industry but has a customer who is in the healthcare industry.

In this example, we will call this company AKYLADE Document Processing Services (ADPS). ADPS is about to take on a few doctor's offices and hospital systems that want to use their services to print out the customer's Explanation of Benefits statements and mail them to the customer's home address. These statements contain protected health information, such as the reason for the doctor's visit and what services were performed on the patient.

Because ADPS will now have access to the patient's electronic medical records to print out these documents and mail them, the company will now be responsible for complying with HIPAA, too.

ADPS has already been using the NIST Cybersecurity Framework for several years, but these are the first clients that work with healthcare-related data. Therefore, ADPS will have to become HIPAA-compliant to conduct this service on behalf of the doctor's offices and hospital systems.

First, ADPS should analyze the HIPAA requirements to see if any new controls must be added to the organization's profile. For example, there is an Administrative Safeguard that requires ADPS to have a "Data Backup Plan" listed under 164.308(a)(7)(ii)(A) in the HIPAA regulation.

The full regulatory text states that an organization shall "[e]stablish and implement procedures to create and maintain retrievable exact copies of electronic protected health information."

Looking at the NIST Cybersecurity Framework, we can identify a related subcategory under PR.DS-11 states, "backups of data are created, protected, maintained, and tested."

ADPS is already following the ISO 27001 standard, as well, so we should cross-reference any of its existing controls that we have in place. For example, ISO 27001 has several requirements related to data backup, including A.12.3.1, A.17.1.2, A.17.1.3, and A.18.1.3.

By carefully reading the ISO requirements and then reviewing my existing data backup plan, which is also ransomware resistant because ADPS has recent and complete offline copies of all of its data, we can see that ADPS is already set up for success and doesn't have to add an entirely new set of controls to meet this HIPAA requirement.

So, by using this 'secure once and comply many' approach, we can ensure that the customer's HIPAA data is backed up using our organization's existing backup plans and methodologies, which already include the products and procedures needed. Since ADPS already uses validated data backup practices that remain consistent with the ISO 27001 requirements, the organization isn't sacrificing compliance with one standard just to comply with the newer HIPAA-based one.

SUMMARY

In this chapter, we explored the concept of profiles within the NIST Cybersecurity Framework.

Profiles provide a customizable approach for organizations to align the framework with their specific goals, risk management strategies, and regulatory compliance requirements. The key components of a profile, including the core functions, categories, and subcategories, serve as the building blocks for defining cybersecurity activities and outcomes.

To create a tailored profile, you should use a step-by-step process to identify organizational requirements and objectives; evaluate the framework core; select and prioritize categories; define subcategories; establish performance goals and metrics; and document and implement the profile.

Profile templates can also be used as a starting point for organizations to create their own profiles, streamlining the profile development process. NIST has even created a handful of sector-specific profiles that can be used as profile templates in certain industries, including the CRI Profile for the Financial Sector, Manufacturing Profile, Election Infrastructure Profile, Satellite Networks Profile, Smart Grid Profile, Connected Vehicle Profiles, Payroll Profile, Maritime Profile, and Communications Profile. These sector-specific profiles provide guidance and tailored recommendations for organizations operating in specific industries, helping them address the unique cybersecurity challenges they may face.

Lastly, the importance of aligning profiles with regulatory compliance requirements was discussed. By leveraging the NIST CSF profiles to support their compliance efforts, organizations can ensure that their cybersecurity practices align with applicable laws, regulations, and industry standards.

Organizations can use profiles to tailor their implementation of the NIST Cybersecurity Framework to their specific needs and objectives, enhance their cybersecurity resilience, and effectively manage their risks. Profiles enable organizations to prioritize and focus their cybersecurity efforts while also supporting compliance requirements within their respective industries.

CHAPTER TEN

ASSESSING CYBERSECURITY RISK

In today's digital landscape, organizations face a wide range of cybersecurity risks that can compromise their operations, assets, and reputation. Assessing these risks and implementing effective risk mitigations are essential steps toward ensuring the security and resilience of an organization's systems and data.

In this chapter, we will apply the concepts covered in the previous chapters through a case study approach. By examining the cybersecurity risk landscape of Certs4U, a fictitious online asynchronous video training company, we will explore how to identify threats, vulnerabilities, and risks and recommend specific risk mitigations to enhance their cybersecurity posture. This practical and real-world example will help illustrate the process of assessing cybersecurity risk and making informed decisions to protect an organization's assets and interests.

CASE STUDY

Certs4U is a leading provider of online asynchronous training programs that specializes in helping students prepare for certification exams across various industries. As an e-learning platform, Certs4U relies heavily on technology infrastructure, data storage, and online communication channels to deliver its training content to students worldwide. With a vast amount of sensitive student data, including personal information and exam results, the company must prioritize cybersecurity and safeguard its systems and data's confidentiality, integrity, and availability.

However, like many organizations operating in the digital space, Certs4U faces numerous cybersecurity risks. Threats such as data breaches, ransomware attacks, and unauthorized access pose significant challenges to the company's operations and reputation. Additionally, vulnerabilities in their technology infrastructure, employee awareness, and third-party dependencies create potential entry points for cyberattacks. It is crucial for Certs4U to conduct a thorough assessment of these threats and vulnerabilities, identify the associated risks, and implement effective risk mitigations to ensure the security and resilience of their operations.

Through the case study of Certs4U, we will delve into the process of evaluating and addressing cybersecurity risks, providing insights into the practical application of risk management principles and strategies.

It is important to note that this case study of Certs4U will serve as an illustrative example throughout this chapter, focusing on a few selected risks and risk mitigations to provide a comprehensive understanding of the cybersecurity risk assessment process, as opposed to providing a full cyber risk assessment of the organization.

IDENTIFYING THREATS

As we dive into the cybersecurity risk assessment process for Certs4U, it is crucial to begin by identifying the potential threats that could jeopardize the organization's systems, data, and operations. Threats encompass a wide range of malicious actors, events, or circumstances that can potentially exploit vulnerabilities and cause harm. By identifying these threats, Certs4U can clearly understand the risks they face and develop appropriate risk mitigations. Some of the threats the organization may encounter include malware and ransomware attacks, social engineering attacks, and attacks by insider threats.

The increasing prevalence of malware and ransomware poses a significant threat to Certs4U's systems and data. Malicious software can infiltrate the organization's networks, compromising the confidentiality and integrity of student information and potentially disrupting their training programs.

Cybercriminals may also attempt to exploit human vulnerabilities within the organization by employing social engineering techniques. Phishing emails, impersonation, and other forms of manipulation can deceive employees into divulging sensitive information or granting unauthorized access to systems.

While Certs4U trusts its employees, there is always a risk of insider threats. An employee with malicious intent or inadvertently negligent behavior can compromise the organization's systems and data, potentially causing significant damage.

IDENTIFYING VULNERABILITIES

In addition to threats, it is crucial to identify vulnerabilities within Certs4U's technology infrastructure, processes, and practices. Vulnerabilities are weaknesses or gaps in security measures that threat actors can exploit. By understanding these vulnerabilities, Certs4U can prioritize its efforts to strengthen its security posture. Some of these vulnerabilities could include inadequate patch management, weak access controls, and insecure third-party dependencies.

If the organization has an inadequate patch management program in place, then it may fail to promptly apply security patches and updates to software and systems, which can leave them vulnerable to known vulnerabilities. Outdated software versions could provide easy entry points for attackers seeking to exploit known vulnerabilities and are a common intrusion vector used by threat actors.

Another vulnerability could be that the organization is using weak access controls. These inadequate access controls and weak passwords could lead to unauthorized access to sensitive student data, accounting systems, or critical information technology systems. Insufficient authentication mechanisms or improper user privileges can also increase the risk of unauthorized access and is a commonly exploited vector used by threat actors.

Certs4U may also rely on numerous third-party vendors or service providers for various aspects of their operations, such as cloud hosting or payment processing. However, these dependencies can introduce vulnerabilities if the third-party organizations do not have robust cybersecurity measures in place. When an organization uses third-party services or software, these still must be considered as part of the organization's overall attack surface. There have been numerous cases where a large organization has suffered a data breach simply because one of its smaller trusted third-party vendors didn't secure its own systems. Once a threat actor is able to exploit the vendor, they can then pivot into the larger organization's network to conduct further exploitation and attacks.

IDENTIFYING RISKS

Having identified the threats and vulnerabilities that Certs4U may encounter, the next step is to analyze and determine the specific risks faced by the organization. By combining the identified threats and vulnerabilities, we can assess the potential impact of these risks on Certs4U's operations and objectives. This process involves evaluating the likelihood of each risk occurring and the potential consequences if they were to materialize.

To effectively identify risks, we will move through combining threats and vulnerabilities to determine which will affect the organization; conduct an assessment of the likelihood of each risk occurring; conduct an impact assessment for each risk; and then perform a risk prioritization to determine which risks should be addressed first.

First, we want to identify all of the threats and vulnerabilities during our initial assessment of the organization, and then we will assess how they intersect and create specific risks. Remember, if there is a threat without a vulnerability, or a vulnerability without a threat, then there is no risk. So, by identifying that the threat of a malware attack exists, and it can be combined with an identified vulnerability of inadequate patch management within the organization, this indicates an increased risk of a successful malware intrusion and a potential data compromise that could occur.

Next, we should determine the likelihood of a risk occurring by evaluating the probability of the associated threat exploiting the identified vulnerability. This assessment considers factors such as the prevalence of similar attacks, the effectiveness of existing security controls, and any historical data or trends relevant to the organization's industry.

After that, we want to assess the potential impact of each risk by considering the potential consequences in terms of financial, operational, reputational, and regulatory aspects. The impact assessment helps prioritize risks based on their potential severity and the magnitude of their potential impact on the organization and its operations.

Then, we should conduct risk prioritization. With a comprehensive understanding of these identified risks, it is essential to prioritize them based on their significance to the organization's objectives. This prioritization ensures that resources and efforts are allocated effectively to mitigate the most critical risks first. Factors considered during prioritization include the likelihood, potential impact, strategic importance, and the organization's risk appetite.

By following these steps, Certs4U can gain a clear understanding of the risks they face and develop an informed risk management strategy.

RECOMMENDATIONS FOR RISK MITIGATION

Based on the identified risks to the organization, it is crucial to propose specific risk mitigation measures that can help minimize the potential impact and likelihood of these risks. A cyber resilience professional should outline targeted strategies to address each identified risk effectively by including risk mitigation measures, including the rationale for the recommended risk mitigation measures, and aligning the recommended mitigations with the organization's objectives and risk appetite.

For each identified risk, tailored risk mitigation measures should be implemented. These measures may include technical controls, process improvements, policy enhancements, and employee awareness and training initiatives. For example, to mitigate the risk of data breaches resulting from inadequate access controls, implementing multi-factor authentication, regular access reviews, and employee cybersecurity training programs can significantly reduce this identified risk.

It is also important to explain the underlying rationale behind each recommended risk mitigation strategy. A cyber resilience professional should highlight the specific benefits and advantages that these risk mitigation measures offer in addressing the identified risks. For instance, implementing regular data backups and disaster recovery mechanisms can mitigate the risk of data loss or system failures, ensuring business continuity and reducing the impact of potential disruptions.

The recommended risk mitigation measures need to be aligned with Certs4U's strategic objectives and risk tolerance. Each recommendation should be evaluated in terms of its feasibility, cost-effectiveness, and potential impact on the organization's operations. It is crucial to consider the organization's risk appetite and ensure that the proposed mitigations strike a balance between enhancing cybersecurity resilience and maintaining the organization's operational efficiency.

By implementing these recommendations, Certs4U can proactively address the identified risks and improve their cybersecurity posture. It is essential to regularly review and update the risk mitigation measures as the threat landscape evolves and new risks emerge.

BENEFITS AND TRADE-OFFS OF RISK MITIGATION

When considering the implementation of recommended risk mitigations for an organization, it is essential to assess the anticipated benefits and understand the trade-offs associated with each strategy. It is also important that you have a solid methodology to aid in your decision-making process.

As you develop your risk mitigation recommendations, it is important to realize that each recommended risk mitigation strategy offers specific benefits that contribute to improving the organization's cybersecurity resilience. These benefits may include enhanced data protection, reduced likelihood of security incidents, improved business continuity, increased customer trust, and regulatory compliance. For example, implementing regular vulnerability assessments and patch management processes can reduce the risk of exploitation by cyber threats, ensuring the integrity and availability of Certs4U's systems and data.

On the other hand, it is important to acknowledge that implementing certain risk mitigations may also involve trade-offs or challenges. These trade-offs can manifest in various ways, such as increased costs, additional resource requirements, changes in user experience, or potential disruptions to existing workflows. For instance, implementing stringent access controls may introduce additional authentication steps, potentially impacting user convenience or locking out legacy systems completely if they cannot technically support the updated authentication process. Therefore, it is crucial to carefully evaluate and balance the potential trade-offs against the anticipated benefits while considering the organization's specific context and priorities.

The decision-making process for selecting or prioritizing specific risk mitigations involves considering factors such as the potential impact on risk reduction, cost-effectiveness, feasibility, and alignment with organizational goals. It requires thoroughly analyzing the benefits, trade-offs, and the organization's risk appetite. A structured approach, such as conducting a cost-benefit analysis or risk assessment, can aid in the decision-making process and ensure that the chosen risk mitigations align with the organization's overall risk management strategy.

By understanding the anticipated benefits and trade-offs of the recommended risk mitigations, Certs4U can make more informed decisions when selecting and prioritizing specific risk mitigation strategies. It is important to involve relevant stakeholders, including executive leadership and management, information technology teams, and the organization's end-users, in the decision-making process to ensure a comprehensive evaluation of the potential impacts and trade-offs associated with each risk mitigation strategy.

EVALUATING THE EFFECTIVENESS OF RISK MITIGATION

Once risk mitigations have been implemented at Certs4U, assessing their effectiveness in reducing risks and achieving the desired outcomes is crucial. This evaluation process involves establishing evaluation criteria, conducting post-implementation assessments, and analyzing the success of the implemented measures.

Clear evaluation criteria needs to be established to evaluate the effectiveness of these risk mitigations. These criteria should be aligned with the objectives set during the risk mitigation planning phase. Common evaluation criteria include:

- The reduction in the likelihood or impact of identified risks
- The improvement of system resilience and availability
- The level of compliance with relevant regulations or standards
- The overall cost-effectiveness of the implemented measures

Post-implementation assessments are then conducted to measure the impact of the implemented risk mitigations and identify gaps or areas requiring further attention. These assessments may involve various methods, such as security audits, vulnerability scans, penetration testing, and incident response exercises. The findings from these assessments provide valuable insights into the implemented measures' effectiveness and help identify areas for potential improvements.

Next, the cyber resilience professional should conduct an analysis of the implemented measures that focuses on evaluating their success in reducing the identified risks. This analysis may involve comparing pre-implementation risk levels with post-implementation levels, evaluating incident trends and patterns, and gathering feedback from various stakeholders. By analyzing these factors, Certs4U will be able to determine the overall effectiveness of the implemented risk mitigations and identify any adjustments or additional measures needed to further reduce risks.

By regularly evaluating the effectiveness of risk mitigations, Certs4U can ensure that the implemented measures continue to address the identified risks and meet their desired objectives. This allows the organization to make informed decisions on refining and improving its risk mitigation strategies by establishing clear evaluation criteria, conducting thorough assessments, and analyzing the success of implemented measures. This process is not necessarily solely linear in nature, though. It is also important to iterate and adapt the risk management approach as new threats emerge, technologies evolve, and the organization's risk landscape changes.

DEVELOPING A RISK MANAGEMENT PLAN

To effectively manage risks, Certs4U will need to develop a comprehensive risk management plan tailored to its specific needs and objectives. The creation of a risk management plan includes:

- Outlining the key components of the plan
- Defining roles and responsibilities within it
- Describing the processes and procedures for ongoing risk monitoring and assessment

The risk management plan for Certs4U should include key components such as risk identification methods, risk assessment criteria, risk treatment strategies, and risk monitoring and review processes. These components provide a structured approach to managing risks and ensure consistency in risk management practices throughout the organization.

Next, it is imperative that the organization clearly defines the roles and responsibilities involved to be able to effectively conduct risk management. The risk management plan should identify the individuals or teams responsible for different aspects of risk management, including risk identification, assessment, treatment, and monitoring. This helps ensure accountability and promotes a shared understanding of everyone's roles in mitigating risks.

Risk management is considered an ongoing process that requires continuous monitoring and assessment. The risk management plan should outline the processes and procedures for regularly monitoring risks, reviewing control effectiveness, and assessing the organization's overall risk posture. This may involve regular risk assessments, vulnerability scans, threat intelligence gathering, and incident reporting mechanisms. By consistently monitoring risks, the organization can identify emerging threats, evaluate the effectiveness of implemented controls, and adapt its risk management approach accordingly.

By developing a robust risk management plan, Certs4U can proactively identify and address risks, protect its assets and operations, and make informed decisions to minimize potential disruptions. The plan serves as a guide for risk management activities, ensuring a systematic and structured approach to addressing risks throughout the organization. It also enables effective communication and coordination among stakeholders, facilitating a shared understanding of the organization's risk management objectives and strategies.

DEVELOPING A CYBERSECURITY STRATEGY

Once the risk management plan has been completed, it is time to develop a comprehensive cybersecurity strategy to effectively protect against cybersecurity threats and ensure a robust security posture within the company. This involves integrating risk management practices into the overall strategic plan, addressing the identified risks and mitigations, and aligning the cybersecurity strategy with the organization's objectives and risk appetite.

Risk management and cybersecurity are closely intertwined, and the cybersecurity strategy should also incorporate risk management practices. By integrating risk assessment, mitigation, and monitoring into the cybersecurity strategy, Certs4U can ensure that cybersecurity measures are aligned with the organization's risk profile and business objectives. This includes identifying high-priority risks, determining the most effective controls, and establishing processes to continuously evaluate and improve cybersecurity measures.

The cybersecurity strategy should specifically address the identified risks and recommended risk mitigations. It outlines the measures and controls that will be implemented to protect against these risks, considering factors such as technology, processes, policies, and people. The strategy should also include clear guidance on implementing and enforcing security controls to mitigate risks effectively.

The cybersecurity strategy being developed needs to align with the organization's overall objectives and risk appetite. This strategy needs to take into account the organization's industry regulations, compliance requirements, and customer expectations. The strategy should also consider cybersecurity incidents' financial, operational, and reputational impacts and align risk management efforts with the organization's risk tolerance.

Certs4U can establish a proactive and holistic approach to cybersecurity by developing a cybersecurity strategy that integrates risk management practices. The strategy provides a roadmap for implementing security measures, addressing identified risks, and aligning cybersecurity efforts with the organization's goals. It also ensures that cybersecurity becomes an integral part of the organization's overall strategic planning process, effectively protecting sensitive data, systems, and operations.

SUMMARY

In this chapter, we looked at a case study focused on a fictional company, Certs4U, to explore the process of assessing cybersecurity risks and recommending risk mitigations. We began by identifying threats and vulnerabilities specific to the organization and thoroughly examining its risks. With a clear understanding of the risks, we proceeded to propose targeted risk mitigation strategies tailored to Certs4U's unique circumstances.

Throughout the chapter, we emphasized the significance of taking a systematic and tailored approach to managing risks.

Organizations can comprehensively understand their risk landscape by combining threat identification, vulnerability assessment, and risk analysis. This enables them to prioritize risks based on their potential impact and likelihood, directing their efforts and resources toward the most critical areas.

Furthermore, we highlighted the ongoing nature of risk management. Cybersecurity risks evolve continuously, making it essential for organizations to establish a continuous monitoring and adaptation culture. The recommended risk mitigation strategies should be regularly assessed for their effectiveness, with adjustments made as necessary to ensure optimal protection.

Following the case study process outlined in this chapter, organizations can strengthen their cybersecurity posture and effectively safeguard their assets. A systematic approach to risk management and tailored risk mitigations enables organizations like Certs4U to mitigate threats, reduce vulnerabilities, and create a more resilient cybersecurity foundation.

PART TWO

THE CR-MAP

CHAPTER ELEVEN

THE CR-MAP PROCESS

Expeditors, a Fortune 500 logistics company, operates as a global freight travel agent, managing freight movements across air, sea, and ground transportation. With a workforce of approximately 18,000 employees, they handle various forms of freight, including ocean, air, road, and rail shipments.

Picture those towering stacks of 40-foot-long steel intermodal shipping containers commonly seen at waterfronts and truck yards. Expeditors are a crucial link in the supply chain for some of its customers, playing a tightly integrated role.

On February 20, 2022, Expeditors made a distressing announcement that its entire global computer network had experienced a complete outage. The outage lasted for three weeks, during which they gradually restored enough functionality to resume serving customers, including basic bookkeeping and accounting tasks. However, it took several additional weeks to fully restore their systems' functionality.

The company openly admitted that the cyberattack it suffered could have a significant negative impact on its financial results. Unfortunately, the impact has been tangible and continues to mount. Expeditors has already incurred losses of $47 million due to disrupted business operations and fines caused by the accumulation of shipping containers in depots and terminals worldwide.

Additionally, they have spent $18 million on investigating and recovering from the incident and covering shipping-related claims. The total cost, which stands at $65 million and continues to rise, is being directly deducted from their cash flow as they did not possess cyber insurance.

In April 2023, one of Expeditors' customers, iRobot Corp., the renowned Roomba robot vacuum cleaner manufacturer, filed a $2.1 million lawsuit against the logistics provider. The lawsuit alleged that Expeditors had breached contractual obligations to ship products and provide real-time inventory data over their 15-year partnership.

iRobot specifically accused Expeditors of inattentiveness and negligence, which exposed their systems to the cyberattack. Furthermore, Expeditors was said to lack a proper business continuity plan to ensure uninterrupted services to iRobot. As a result, iRobot had to physically count their products at Expeditors' warehouses and arrange to load nearly 12,000 pallets into 207 rented tractor-trailers to fulfill customer orders. The delayed deliveries forced iRobot to refund retailers an amount of $900,000 and spend an additional $23,000 on expedited shipping to reach consumers directly.

According to their contract, Expeditors was obligated to receive iRobot's new products, store and maintain them, and ship them to customers within 24 hours of order receipt. Additionally, Expeditors had to update their system within four hours of any order or stock movement. However, when Expeditors shut down its operating systems, all the services iRobot relied on suddenly stopped. Products in transit were left idle, and customer orders went unfulfilled.

To fulfill their customer commitments, iRobot had to switch to a new logistics provider, incurring an additional cost of $1.1 million.

Despite their best efforts to mitigate the effects of Expeditors' failure, iRobot faced an extra $1 million in storage costs and back charges from retailers.

iRobot claims that Expeditors' 10-K filing had identified the cyberattack as a foreseeable risk. Yet, the company failed to implement any risk mitigation measures. As a result, iRobot is seeking a minimum compensation of $2.1 million, along with interest at the statutory rate of 9% from the breach date, as well as court costs and legal fees.

While the outcome of the lawsuit may remain undisclosed due to potential out-of-court settlements, the public release of the details underscores the severe financial consequences Expeditors has faced due to their lack of cyber resilience.

REASONABLE AND REPEATABLE CYBERSECURITY PRACTICES

As we have witnessed, the magnitude of cyber threats continues to rise. It is projected to increase further in the years ahead. Recognizing this new reality, the United States government introduced the NIST Cybersecurity Framework, emphasizing the need for reasonable cybersecurity practices rather than solely relying on prevention measures. This framework was developed as a response to the escalating cyber threat landscape. It provided organizations with a flexible and adaptable approach to effectively manage cyber risks. As either a cyber resilience professional or an executive leader, it is crucial to understand and adhere to this new standard in dealing with cyber risks, as formidable forces are at play.

On one side, adversaries persistently launch cyber attacks, making it inevitable that your organization will experience a breach at some point. Therefore, alongside your ongoing prevention efforts, preparing for the eventuality of a breach without prior warning is essential, as your public reputation is at stake.

To emphasize the human factor in cybersecurity, it is imperative to cultivate a culture within your organization that values daily cyber hygiene and the practice of reasonable cybersecurity. This includes raising employee awareness, providing regular training programs, and fostering a sense of responsibility among all staff members to actively participate in protecting the organization's digital assets. Organizations can significantly strengthen their defense against cyber threats by developing a cyber-resilient culture.

On the other side, the Federal Trade Commission (FTC) mandates that organizations implement "reasonable security measures" based on their size, sophistication, and data collection practices. Failure to comply may result in charges of unfair or deceptive acts, leading to severe consequences such as corrective orders, extensive oversight of cybersecurity programs for up to two decades, and fines of up to $40,000 per violation.

What follows is not just a theoretical concept or hypothetical scenario; it is a proven, repeatable system that has helped numerous organizations around the globe to get a handle on their organization's cybersecurity risks. As the threat landscape continues to evolve, it is no longer enough to focus solely on good cyber hygiene and defense.

Instead, your organization must demonstrate that it uses reasonable cybersecurity practices and has a repeatable approach to maintaining its cyber resiliency. Documentation and implementation are critical, as they provide evidence of your cybersecurity practices, especially in situations such as potential acquisitions or investigations.

Furthermore, it is imperative to cultivate a culture within your organization that values daily cyber hygiene and the practice of reasonable cybersecurity. You can thrive amidst evolving cyber risks by gaining insights from the front lines and using that information to drive necessary changes. As noted by leadership expert John P. Kotter, effective leadership lies in facilitating meaningful change.

Within your grasp, you hold a practical guide that will assist you in implementing and documenting your cybersecurity plan to such a degree that it not only safeguards against reasonable threats, but also protects you during potential acquisitions and investigations.

THREE PHASES OF THE CYBER RISK MANAGEMENT ACTION PLAN

To establish a robust cyber resilience program within your organization, you should consider implementing the Cyber Risk Management Action Plan (CR-MAP) process. This process will significantly enhance your organization's cybersecurity while positively impacting your bottom line and saving substantial time.

This comprehensive plan encompasses three distinct phases and serves as a practical roadmap to managing cyber risks effectively. These phases include actions to discover your top cyber risks, making your Cyber Risk Management Action Plan (CR-MAP), and performing maintenance and updates.

The first phase of the CR-MAP (Phase One) is to discover your top cyber risks. This phase spans a period of thirty days. It focuses on assessing and evaluating your company's existing cyber risks. As an executive, you encounter infinite cyber risks that can impact your organization, but it is crucial to prioritize them effectively, given your limited resources. During this phase, your goal is to develop a rigorous prioritization method to determine your company's most critical risks.

The second phase of the CR-MAP (Phase Two) is used to make your actual Cyber Risk Management Action Plan. This phase also lasts thirty days and involves the creation of a personalized cyber risk management strategy tailored to your organization's specific needs. This strategy should address the top five cyber risks identified in the previous phase. Throughout this phase, you will ensure that every dollar invested in cyber risk management provides maximum value in mitigating one of the prioritized risks. Your strategy will encompass four key dimensions of business value: technical risk mitigation, enhanced operational reliability, legal risk mitigation, and financial returns. By adopting reasonable cybersecurity practices, you will manage your risks effectively and improve your competitive advantage.

The third phase of the CR-MAP (Phase Three) is used to conduct maintenance and updates to the organization's cybersecurity posture and its Cyber Risk Management Action Plan. This phase spans a period of at least ten months, and when combined with Phase One and Phase Two of the CR-MAP process, it completes the first-year cycle of the three-phase plan. Cybersecurity is an ongoing journey, and Phase Three of the CR-MAP emphasizes the importance of implementing, maintaining, and continuously improving your organization's risk management strategy.

Throughout Phase Three, regular monthly check-ins and comprehensive quarterly reviews will be conducted to monitor your organization's progress. If any gaps or challenges arise in implementing the Cyber Risk Management Action Plan effectively, this maintenance phase offers an opportunity to investigate the underlying reasons and develop corrective measures to regain momentum. Additionally, it serves as a platform to acknowledge and celebrate your company's achievements in managing cyber risks.

By following this comprehensive approach and committing to ongoing vigilance, you can bolster your organization's cyber resilience and ensure effective management of cyber risks in a rapidly evolving threat landscape.

THE THIRTEENTH MONTH

After successfully concluding the third phase, which signifies the completion of an entire year (or more) since commencing the first phase of the CR-MAP, you might question whether the journey has come to an end.

What awaits you beyond this point?

The answer is straightforward: you must return to the first phase and repeat the three-phase CR-MAP process again. Time after time, you will persistently strive for improvement and progress.

It is important to note that the progress made during the first year does not go to waste. On the contrary, based on the Cyber Risk Management Action Plan you have developed, more work is likely needed to refine and enhance your cyber risk management efforts. It would be a misstep to simply halt the process at this stage and claim victory.

After all, cyber risks continually evolve as adversaries innovate, and your business can also undergo drastic changes over time. Additionally, the interview process and discussions surrounding cyber risks serve as valuable reminders for your employees regarding the significance of practicing good cyber hygiene and maintaining reasonable cybersecurity practices.

Therefore, it is prudent to embrace the structured, systematic, and comprehensive approach provided by the Cyber Risk Management Action Plan on an annual basis within your organization. Repeating the three phases each year ensures that your organization remains resilient and adaptive to evolving cyber threats while continuously improving your cyber risk management capabilities.

THE FIVE QUESTIONS

Once you have created your Cyber Risk Management Action Plan (CR-MAP), you should be able to confidently answer the following five questions:

1. What are the top five cyber risks to my organization?

2. Am I getting the biggest return possible for my cyber risk management dollars?

3. Do all our organization's executives and leaders understand our cybersecurity plans?

4. Does everyone at work know how they can help to mitigate our top cyber risks?

5. What do I tell our biggest customers or stakeholders when they ask, "What are you all doing about cybersecurity?"

These five questions serve as a crucial checkpoint for evaluating the effectiveness of your organization's Cyber Risk Management Action Plan. By confidently answering these questions, you demonstrate a strong understanding of your top cyber risks, the return on investment for your cyber risk management efforts, the level of awareness and understanding among executives and leaders, the engagement of all employees in mitigating cyber risks, and your ability to communicate your cybersecurity measures to key stakeholders.

These questions are used to guide you in assessing the comprehensiveness and alignment of your cybersecurity practices, ensuring that your organization is well-prepared to navigate the evolving threat landscape and meet the expectations of its customers and stakeholders. Continually revisiting and refining your responses to these questions can enhance your organization's cyber resilience and maintain a proactive and informed approach to cybersecurity.

ATTORNEY-CLIENT PRIVILEGE

Trying to actively manage every cyber risk facing your company can be overwhelming. However, even by working through the Cyber Risk Management Action Plan process and identifying your risks, you will already be ahead of most other organizations in terms of cyber risk management.

Before delving into the process, it is crucial to understand the concept of attorney-client privilege and its potential role in protecting yourself and your organization. **Attorney-client privilege** is a legal concept that protects communications between a client and their attorney from being disclosed to third parties to ensure confidentiality of information shared during legal advice or representation. This concept is protected under the Federal Rules of Evidence, common law, and is also codified in some state statutes in the United States. This same concept may apply to many other countries around the world, but you should consult with an attorney in your location to verify that this privilege exists in your particular country. Of note, this privilege does have some limited exceptions and may not apply in all instances and circumstances. Again, be sure to consult a licensed attorney for specifics.

If one of your identified cyber risks manifests itself as a data breach, failure to prioritize that risk in your Cyber Risk Management Action Plan could lead to potential liability as well as accusations of negligence, both legally and in the court of public opinion. By conducting this work under attorney-client privilege, you may be able to retain control over your cyber risk records. Additionally, an attorney can help you navigate any potential requests for a copy of those records as potential evidence.

It is important to note that establishing attorney-client privilege over your cyber risk management records may also come with certain disadvantages, such as increased costs and potentially longer timelines to complete the work. That said and based upon over two decades of commercial and business litigation experience, one of our authors can confirm that it is almost always less expensive to retain proper legal advice from a licensed attorney up front to mitigate your liability exposure than to hire a lawyer to defend your organization against litigation later on.

Please note that this book is not to be considered legal advice. Instead, we recommend that you check with your organization's in-house counsel (if it has one). If your organization does not have in-house counsel, you should consider hiring an outside attorney with cybersecurity expertise to assist your organization in establishing an attorney-client relationship with privileged communications, as well as developing best practices for your organization's cyber risk management activities.

Licensed counsel can assist with preparing contractual arrangements for your organization's work with a qualified cyber resilience practitioner who would then follow your attorney's instructions to ensure that the practitioner does not improperly handle or share your organization's cyber risk information. Including an attorney in the retention and instruction of a cyber resilience practitioner for your organization may also extend the attorney-client privilege to the development of your organization's Cyber Risk Management Action Plan (CR-MAP).

By seeking to incorporate attorney-client privilege into your cyber risk management practices, you can add an extra layer of protection and maintain more control over your cyber risk records. This will bolster your organization's ability to respond effectively to potential legal challenges and safeguard your reputation in the face of cyber incidents.

SUMMARY

This chapter introduced the three-phase Cyber Risk Management Action Plan (CR-MAP) process. The first phase focuses on discovering and assessing the top cyber risks faced by the organization. This phase involves prioritizing risks to allocate limited resources effectively. The second phase involves creating a personalized Cyber Risk Management Action Plan (CR-MAP) that addresses the top five risks identified in Phase One. This plan considers various dimensions of business value and aims to mitigate risks while enhancing operational reliability and competitive advantage. The third phase emphasizes the importance of maintaining, updating, and continuously improving the organization's cybersecurity posture and risk management strategy.

The concept of the thirteenth month was introduced to emphasize the need for an ongoing and iterative approach to cyber risk management. Rather than considering the completion of the initial three-phase cycle as the end of the journey, organizations are encouraged to repeat the CR-MAP process annually. This repetition ensures continual improvement and adaptation to evolving cyber threats, as well as the alignment of cyber risk management efforts with changing business dynamics.

In summary, the CR-MAP process provides organizations with a structured and comprehensive framework to strengthen their cyber resilience. By systematically identifying and managing cyber risks, organizations can proactively mitigate potential incidents, safeguard valuable assets, and sustain a competitive advantage in the modern digital landscape. However, it is vital to emphasize the significance of perpetual vigilance, continuous improvement, and the ingrained integration of cybersecurity practices into the organization's culture to effectively achieve these objectives.

CHAPTER TWELVE

PHASE ONE: DISCOVERING TOP CYBER RISKS

Phase One, which takes place over the course of thirty days, is comprised of measuring and scoring your company's current cyber risks. As an executive, you encounter unlimited risks to your company, but your resources to manage those risks are limited, so you need a strict prioritization method. That's what you'll develop in Phase One: priority.

The first phase of the Cyber Risk Management Action Plan (CR-MAP) spans 30 days. It marks the critical stage of measuring and scoring your company's existing cyber risks.

As a cyber resilience professional, you should be well aware of the countless risks that can impact an organization. However, the resources available to manage these risks are limited; thus, a rigorous prioritization method is required to determine which risks will receive which resources to help mitigate, transfer, avoid, or accept those identified risks.

During the first phase of the CR-MAP process, our goal is "Discovering Top Cyber Risks" within the organization. This will help us as we embark on the journey of establishing clear priorities for our cyber risk management strategy. By the end of this phase, we will have a comprehensive understanding of the most significant risks an organization faces and be equipped with the necessary insights to allocate organizational resources effectively.

The first phase of the Cyber Risk Management Action Plan includes eight steps:

1. Widen Your Scope
2. Get Buy-In
3. Select Interviewees
4. Generate the Questionnaire
5. Determine Your Target Scores
6. Conduct the Interviews
7. Compile and Average the Scores
8. Communicate Your Top Five Cyber Risks

Let us delve into the intricacies of Phase One and explore how this foundational phase and its eight steps can set the stage for a resilient and proactive approach to managing cyber risks.

STEP 1: WIDEN YOUR SCOPE

Before laying the foundation of your Cyber Risk Management Action Plan (CR-MAP), it is crucial to broaden the scope of what you want to measure and evaluate within the organization. When guiding stakeholders through the CR-MAP process, we often find that they have a narrow focus primarily centered on improving their technological defenses. They may prioritize protecting customer credit card information or preventing the leakage of user password data while completely overlooking other key and important areas of risk. While these technical defenses are important considerations, we always encourage organizations to expand their scope and measure every facet of their cybersecurity risk posture when embarking on creating a Cyber Risk Management Action Plan.

This broader perspective encompasses a comprehensive assessment of the people, processes, management, and technology used by the organization to achieve its mission. The people perspective focuses on the organization's employees and their awareness and adherence to cybersecurity practices. The process perspective is used to evaluate the effectiveness of established procedures and protocols. The management perspective is focused on the organizational leadership's commitment and oversight of cybersecurity initiatives. The technology perspective assesses the robustness of the technological systems and infrastructure.

It is essential to recognize that cybersecurity encompasses more than just technological aspects. An organization must include all of its digital assets, such as customer data, payroll data, reputation, and trade secrets, within the purview of its assessment. Moreover, involving all departments across the organization is crucial, ensuring a holistic understanding and involvement in cybersecurity efforts.

We recommend incorporating all 22 categories outlined in the NIST Cybersecurity Framework to guide your assessment. While this assessment is not an external audit or a response to regulatory requirements, it is a conscious choice you make to enhance your cybersecurity. Therefore, embracing a more comprehensive approach is prudent as you begin developing your Cyber Risk Management Action Plan (CR-MAP).

Transparency and open communication are paramount. If any information is hidden or withheld, it will only hinder your progress and jeopardize the effectiveness of your cybersecurity initiatives. By adopting this inclusive approach and thoroughly examining all facets of your organization's cybersecurity, you demonstrate a commitment to comprehensive protection and lay the groundwork for a robust CR-MAP.

FUNCTION	CATEGORY	CATEGORY IDENTIFIER
Govern (GV)	Organizational Context	GV.OC
	Risk Management Strategy	GV.RM
	Roles, Responsibilities, and Authorities	GV.RR
	Policy	GV.PO
	Oversight	GV.OV
	Cybersecurity Supply Chain Risk Management	GV.SC
Identify (ID)	Asset Management	ID.AM
	Risk Assessment	ID.RA
	Improvement	ID.IM
Protect (PR)	Identity, Management, Authentication, and Access Control	PR.AA
	Awareness and Training	PR.AT
	Data Security	PR.DS
	Platform Security	PR.PS
	Technology Infrastructure Resilience	PR.IR
Detect (DE)	Continuous Monitoring	DE.CM
	Adverse Event Analysis	DE.AE
Respond (RS)	Incident Management	RS.MA
	Incident Analysis	RS.AN
	Incident Response Reporting and Communication	RS.CO
	Incident Mitigation	RS.MI
Recover (RC)	Incident Recovery Plan Execution	RC.RP
	Incident Recovery Communication	RC.CO

STEP 2: GET BUY-IN

This Cyber Risk Management Action Plan you are developing simply will not work without buy-in from your employees. Gathering this buy-in begins with the tone and approach you adopt when engaging with the organization's stakeholders. Your ability to foster a collaborative mindset is essential when you distribute questionnaires and seek open and honest input on their perceptions and activities within the company. The way you communicate about the organization's new action plan will set the stage for a more collaborative approach.

You will want to conduct an assessment, not an audit. Some people use the terms audit and assessment synonymously, but we don't. To us, they're strikingly dissimilar.

An **audit** involves an external evaluation aimed at finding faults within the organization, most insiders assume. An audit generally puts people on the defensive.

Assessments, however, are different. **Assessments** are internal management actions focused on identifying areas for improvement. Even if an outsider assists with the assessment, the assessment process itself remains owned by management, is internally focused, and is focused on a different objective than audits.

When reaching out to the organization's employees for the first time, having an internal champion assist with getting the initial message out to the potential interviewees can be helpful. This is extremely helpful if you are being brought in as part of an outside consulting team that doesn't already have strong relationships within the organization or with its affected stakeholders.

In our experience, having someone from your organization's executive management or C-Suite (CEO, COO, CFO, CIO, or CISO) send out the initial interview selection email can effectively gain initial support. On the following page, for your reference and review, is a sample email template the organization's leadership can use.

It is important to ensure the team knows you're neither here as an adversary, nor here to cause them issues. Instead, you are simply asking questions in order to gather valuable information. To that end, you should be open about the process your company is undertaking, and your email communication should reflect that open, collaborative spirit. You will inevitably get questions in return, many of which will spawn from people's anxieties about being interviewed. Be sure that your response to those questions reinforces the collaborative nature of the interview.

Sample Email Template

To: All Hands
From: CEO
Re: Cyber Risk Assessment

Hello Team,

I am extremely concerned about our cybersecurity, and I hope you are, too. The crooks are out there, and we can't even begin to imagine how they are penetrating companies, stealing trade secrets, money, and customer lists, and generally disrupting businesses.

In order to combat this, we are implementing a Cyber Risk Management Action Plan. Our first step in implementing that plan is to conduct interviews with many of you so that we can fully assess our current cyber risks.

These interviews aim to learn how we can best balance our cybersecurity needs with our day-to-day business needs. Each interview will take 30–90 minutes and will be conducted in person here in [city] or via a video call, as necessary.

[Point of contact name] will coordinate with each of you to find a workable time slot for your interview.

If you have any questions about the program itself, please let me know.

If you have questions about the logistics of the interviews, please ask [point of contact name] for assistance.

Thank you!
[Your name]

STEP 3: SELECT INTERVIEWEES

During the first phase of the Cyber Risk Management Action Plan, you must determine the right individuals to interview during your assessment. The number and type of interviews conducted will depend on the size and revenue of your company.

As a general guideline, conducting fifteen to twenty in-person interviews is recommended if your company has an annual revenue of approximately $100 million. If your annual revenue is less than $10 million, forming a group of around six individuals to generate scores is considered a reasonable approach.

If you are working with a larger organization with revenues of over $1 billion annually, sending out electronic questionnaires instead of conducting in-person interviews may be a more cost-effective option. Online platforms like Survey Monkey or Google Forms can be utilized for administering questionnaires in an asynchronous and passive manner. As a reminder, be sure to use appropriate security measures if using a third-party program to avoid inadvertent public disclosure of your company's evaluation process and data.

Alternatively, if you are working with a non-profit or any other organization where the ratio of revenue to interviews would drastically skew the number of interviews you might conduct, you can use a different guideline to select the appropriate number of employees to interview.

For example, we have worked with several non-profits where we have used guidelines based on a percentage of their total staff size instead of relying on the revenue targets listed above. In these cases, we have found that an interview ratio of 1:5 is appropriate if the organization has 50 or less staff members. If the organization has between 51-1000 staff members, then a ratio of 1:10 is more appropriate. As you begin to work with larger organizations, you should continue to increase the ratio, thereby decreasing the overall number of interviews needing to be conducted. Again, these are loose guidelines that you can adjust based on your target organization's needs and specific use case.

When selecting interviewees, it is important to focus on middle managers and senior-level individual contributors from key departments such as finance, human resources, operations, legal, and information technology. These individuals, whom we will refer to as the organization's cyber risk influencers, possess valuable knowledge of the cyber risk practices occurring at the operational level. While they may not be cybersecurity experts, they offer firsthand insight into a given organization's cyber risk landscape. Additionally, the very act of interviewing them will make them better cyber risk managers, a trait that will shape other members of your workforce and will shift your organization's culture toward more reasonable cybersecurity.

We have found these middle managers and senior-level individual contributors to be at the sweet spot for providing us with all kinds of valuable information during the interview process. If you instead choose more junior-level employees, they might lack the necessary perspective to provide substantial input. On the other hand, if you interview only senior executives, they are often more detached from day-to-day operations and don't fully understand the operational cyber risks that exist within the organization.

If your organization operates across multiple geographic locations, it is essential to also include influencers from different regions to capture a more comprehensive understanding of the organization's complete cyber risk landscape. By selecting the right individuals for interviews, you ensure that you gather insights from those who possess relevant knowledge and can contribute meaningful perspectives to developing the organization's new Cyber Risk Management Action Plan.

STEP 4: GENERATE THE QUESTIONNAIRE

The questionnaire administered during the interviews plays a pivotal role in collecting crucial data for your Cyber Risk Management Action Plan. Through a series of carefully crafted questions tailored to align with the NIST Cybersecurity Framework, you will assess your organization's adherence to key cybersecurity principles, which will increase your cyber resilience.

These questions serve as an essential tool for evaluating your organization's capabilities in governing, identifying, protecting against, detecting, responding to, and recovering from cybersecurity risks. This is critically important to any organizations that are based in the United States because there are now laws that require organizations to meet certain levels of cyber resiliency and cybersecurity.

⸻ For example, the Federal Trade Commission (FTC) Safeguards Rule of 2023 requires non-banking financial institutions to develop, deploy, and maintain a comprehensive security program to keep customer financial data safe. One way to meet this requirement is to implement the NIST Cybersecurity Framework and your associated Cyber Risk Management Action Plan. To ensure you have a comprehensive plan in place, you should consider the following six questions:

1. How well does the organization <u>govern</u> its risk management strategy, expectations, and policy?

2. How well does the organization <u>identify</u> digital assets and cyber risks?

3. How well does the organization <u>protect</u> your assets against those risks?

4. How well does the organization <u>detect</u> cybersecurity breaches?

5. How well does the organization <u>respond</u> to those breaches?

6. How well does the organization <u>recover</u> from those breaches?

Notice that the six underlined keywords in these questions all directly map back to the six functions presented in the NIST Cybersecurity Framework. By leveraging this basic questionnaire, you will gain valuable insights into the effectiveness of your cybersecurity practices and uncover areas for improvement, ultimately strengthening your organization's overall cyber resilience.

When you create your questionnaire, you can keep it broad using just these six questions and ask the interviewees to assign a value from zero through ten to each response. We'll further explain this scoring scale below.

For other engagements, you may find that these questions don't dig deep enough to get the answers you seek. In these cases, you will want to use a more in-depth series of questions to generate the necessary responses from your interviewees.

Most people believe you can never have too many resources dedicated to protecting a digital asset in your organization, but this is simply not true. It is possible to have too much (or too little) security. Your risk mitigations and controls should be based on a reasonable level of protection based on the asset being protected.

For example, building a $100,000 state-of-the-art garage to house a $5,000 car would be ridiculous. Not only is the new garage too expensive and complicated for the relatively low-valued car, but it is also really unnecessary, given the asset value we are trying to protect. In this case, I could have the $5,000 car destroyed twenty times before the added cost of protecting it using this new garage would have made financial sense.

Similarly, we don't want to overengineer a security solution to protect a relatively low-value asset. In addition to wasting money, we will also be adding a lot of friction to the organization's workflows and processes if we don't think holistically about the controls being recommended for inclusion as part of the Cyber Risk Management Action Plan.

There is always a tradeoff between security and operations. As we make things more secure, we often reduce our operational efficiencies. In fact, an excess of cybersecurity controls can hinder productivity, leading individuals to seek alternative methods to complete their assigned tasks.

If we require that every system utilized by the organization has a different long, strong, and complex password, then the end user will be confronted with an overwhelming number of user IDs and passwords to remember and manage. Most people simply can't, or won't, remember them all. Therefore, they resort to writing down the passwords in a notebook or sticky note. However, this defeats the entire purpose of long, strong, and complex passwords because the note can now easily be stolen, and the password discovered.

As each question is asked, the interviewee will be required to provide a score from zero through ten.

If the score is between zero and four, this is generally considered to represent a level of insecurity, from no security at all to some security.

If the score is between five and eight, this is generally considered to provide an acceptable level of security, from minimally acceptable security to fully optimized security.

If the score is either a nine or a ten, this is generally considered to provide too much security and a waste of resources, including time, money, and employee morale.

We have created a standard score key to use during the interview process that uses definitive statements about the organization's cybersecurity practices to better identify the appropriate number for an interviewee to use.

So, how do you implement this in the real world?

Let's pretend you have been brought in as an external consultant to assist with running an organization-wide self-assessment. You are interviewing one of the organization's identified cyber risk influencers and ask, "How well has the organization established and implemented the processes to identify, assess, and manage supply chain cybersecurity risks?"

YOUR EVALUATION	SCORE
Our organization rarely or never does this.	0
Our organization sometimes does this, but unreliably, and re-work is common.	3
Our organization does this consistently with some minor flaws from time to time.	5
Our organization does this consistently with great effectiveness and high quality.	8
Our organization does this at excessive financial cost, and people can't easily get their work done.	10
I'm not sure if our organization does this or not.	UNK
This is not applicable to us.	N/A

Using the potential scorecard, the cyber risk influencer states, "Our organization rarely or never does this," which corresponds with a score of zero. If your influencer knows or perceives that the organization is a bit better than that, then they could have instead read the next score statement,

"Our organization sometimes does this, but unreliably, and re-work is common." A score of three is recorded as the answer to this question.

If your influencer knows or perceives that your organization is better than a three, then they read the next score statement, and so on, until they find a statement that most closely matches their perception from the associated scoring statements from zero to ten.

It is important to recognize that there are two other possible responses that the influencer can choose: unknown and not applicable. Usually, these responses should be seldom used if you have properly identified the right cyber risk influencers to interview, but they occasionally happen. For example, if you are interviewing a senior database administrator and ask them a question about how accounting does a certain process, then an unknown answer would be appropriate. But, if you are asking if the database is using data-at-rest encryption and the database administrator tells you that it is unknown, this may be a clue that you do not have the right cyber risk influencer involved in your interview process, and you may need to ask for another influencer to find out the true nature of the organization's cyber risk.

Now, you might be wondering how you'll develop the exact questions to use in your questionnaire. While you could write a question for each of the 106 subcategories in the NIST Cybersecurity Framework, that might be unnecessary for your use case. If you are working as a consultant working with the NIST Cybersecurity Framework on a daily basis, though, you may prepare a set of questions for each of the subcategories and reuse them across many of your engagements.

However lengthy it is, your questionnaire should be tailored to the specific organization that you are working with. In most of our engagements, we focus on 24 specific questions that have provided us with the best information about an organization's current cyber risk posture. You can use these questions as a baseline for your default questionnaire; however, the below questionnaire is malleable and can be molded based upon your experience and needs, allowing you to add useful questions or remove unneeded ones.

In the enumerated list below, you will find the question, each activity as identified by their function and category, and some clarifying notes about each question. Additionally, some of these categories involve subparts which will be indicated with a bracket and a letter, such as [a] or [b]. For example, GV.SC[b] relates to the Govern function, the Cybersecurity Supply Chain Risk Management category, and the involvement of organizational stakeholders; whereas, GV.SC[a] refers to the Govern function, the Cybersecurity Supply Chain Risk Management category, and the physical and digital assets.

Please note that while these questions have been numbered in this textbook for clarity, they can be used in any order you deem appropriate.

1. How well does your organization understand its larger operating context when making cybersecurity risk management decisions?
    - Organizational Context (GV.OC)
    - The term context used in this question refers to things such as the organization's mission; internal and external stakeholder expectations; critical dependencies; and legal, regulatory, and contractual requirements.

2. How well does your organization establish and use priorities, constraints, risk tolerance, and risk appetite when making operational risk decisions?
    - Risk Management Strategy (GV.RM)
    - Risk tolerance refers to the amount of a loss that an organization is willing to experience given their existing assets and the other risks they currently face. Risk appetite refers to the amount of risk that the organization is comfortable accepting.

3. How well does your organization establish and communicate cybersecurity roles, responsibilities, and authorities to foster accountability, performance assessment, and continuous improvement?
    - Roles, Responsibilities, and Authorities (GV.RR)
    - You should consider accountability, performance assessment, and continuous improvement independently within the cybersecurity context of the organization.

4. How well does your organization establish, communicate, and enforce organizational cybersecurity policy?
   o   Policy (GV.PO)
   o   The organization's policy may be written out formally, or simply shared verbally pursuant to an informal policy.

5. How well does your organization use the results of organization-wide cybersecurity risk management activities and performance to inform, improve, and adjust the risk management strategy?
   o   Oversight (GV.OV)
   o   When considering the results, you should ensure that the organization has a continuous feedback loop in place so that continuous improvement is occurring within the organization.

6. How well does your organization identify, establish, manage, monitor, and improve cybersecurity supply chain risk management processes?
   o   Cybersecurity Supply Chain Risk Management (GV.SC[a])
   o   Cybersecurity supply chain refers equally to both digital and physical assets for the purposes of this question.

7. How well does your organization involve organizational stakeholders when identifying, establishing, managing, monitoring, and improving cyber supply chain risk management processes?
   o   Cybersecurity Supply Chain Risk Management (GV.SC[b])
   o   Organizational stakeholders refers to both internal and external groups and organizations that have a significant interest in your organization's well-being for the purpose of this question.

8. How well does your organization identify and manage assets that enable the organization to achieve business purposes consistent with their relative importance to organizational objectives and the organization's risk strategy?
   o   Asset Management (ID.AM)
   o   In the NIST Cybersecurity Framework, assets are broadly defined to include data, hardware, software, systems, facilities, services, and people.

9. How well does your organization understand the cybersecurity risk to the organization, assets, and individuals?
   - Risk Assessment (ID.RA)
   - The organization's understanding of cybersecurity risk should be measured by the thoroughness of their risk management documentation, the training level of their staff, and the effectiveness of the staff applying that knowledge to the organization's cybersecurity risk decisions.

10. How well does your organization identify improvements to organizational cybersecurity risk management processes, procedures, and activities across all NIST Cybersecurity Framework functions?
    - Improvement (ID.IM)
    - It is important to consider all six functions, including Govern, Identify, Protect, Detect, Respond, and Recover.

11. How well does your organization limit physical and logical access of authorized users, services, and hardware to assets?
    - Identity Management, Authentication, and Access Control (PR.AA[a])
    - The organization should implement the principle of least privilege, which refers to providing a user with the least amount of access necessary for them to perform their job functions or role.
    - This also applies to assets, including systems, services, and hardware devices, such as IoT devices.

12. How well does your organization manage physical and logical access relative to the organization's assessed risk of unauthorized access?
    - Identity Management, Authentication, and Access Control (PR.AA[b])
    - If the value of a given asset is higher, your organization should implement more rigorous methods to manage its users' physical and logical access in order to prevent unauthorized access.
    - Conversely, if the value of a given asset is lower, your organization may opt to implement less rigorous methods.

13. How well does your organization provide the organization's personnel with cybersecurity awareness and training so that they can perform their cybersecurity-related tasks?
    o   Awareness and Training (PR.AT)
    o   The organization's staff should be trained in the fundamentals of cybersecurity so that they can identify potential threats and vulnerabilities.
    o   All users within the organization that handle sensitive data and systems need to be trained to prevent becoming the victim of a social engineering-based attack.
    o   This activity also supports the Roles, Responsibilities, and Authorities (GV.RR) activity.

14. How well does your organization manage its data based on the organization's risk strategy to protect the confidentiality, integrity, and availability of its information?
    o   Data Security (PR.DS)
    o   Confidentiality, integrity, and availability are the foundational concepts that must be considered in order to protect an organization's data.
    o   The organization's users should practice good cyber hygiene to prevent a cyber security incident.
    o   Good cyber hygiene includes using a high-quality password manager, encrypting data during storage and transfer, and following appropriate data protection checklists to prevent potential data loss and leakage.

15. How well does your organization manage the hardware, software, and services of physical and virtual platforms based on the organization's risk strategy to protect their confidentiality, integrity, and availability?
    o   Platform Security (PR.PS)
    o   Confidentiality, integrity, and availability are the foundational concepts that must be considered in order to protect an organization's data.
    o   The organization's system administrators should ensure that the physical and virtual platforms are secure, patched against known vulnerabilities, and protected against known threats.

16. How well does your organization manage security architectures based on the organization's risk strategy to protect asset confidentiality, integrity, and availability, as well as organizational resilience?
    - Technology Infrastructure Resilience (PR.IR)
    - The organization should identify and mitigate any single point of failure in its technology stack.
    - The organization should also ensure its security architecture is well-documented and understood by its system administrators.

17. How well does your organization monitor assets to find anomalies, indicators of compromise, and other potentially adverse events?
    - Continuous Monitoring (DE.CM)
    - The organization should be conducting continuous monitoring using a variety of methods, such as system event log collection, network event log collection, log centralization, network egress filtering, and monitoring honeypots.

18. How well does your organization analyze anomalies, indicators of compromise, and other potentially adverse events to characterize the events and detect cybersecurity incidents?
    - Adverse Event Analysis (DE.AE)
    - The organization should be utilizing monitoring and detection processes to identify potentially adverse cyber events. The organization can achieve this by analyzing the organization's system event logs and network event logs, centralizing log collection and analysis, conducting network egress filtering, and monitoring honeypots.

19. How well does your organization manage responses to detected cybersecurity incidents?
    - Incident Management (RS.MA)
    - The organization should be able to effectively detect and respond to potentially adverse cyber events.
    - Some ways for the organization to achieve this are by prioritizing incidents based on their scope, understanding their likely impact, and considering the time-critical nature of each event.

20. How well does your organization conduct investigations that implement effective response actions, support forensic activities, and prepare for their future recovery efforts?

    o   Incident Analysis (RS.AN)
    o   The organization's incident coordinator should document the incident in detail during an investigation.
    o   They are also responsible for preserving both the integrity of the documentation and the sources of all information being reported.

21. How well does your organization coordinate response activities with internal and external stakeholders as required by laws, regulations, or policies?
    o   Incident Response Reporting and Communication (RS.CO)
    o   The organization should have a credible and tested plan that adequately coordinates its crisis communication methods between the organization and its internal or external stakeholders.

22. How well does your organization perform activities to prevent the expansion of an event and mitigate its effects?
    o   Incident Mitigation (RS.MI)
    o   The organization's mitigation activities may include the use of cybersecurity technologies (e.g., antivirus software) and cybersecurity features of other technologies (e.g., operating systems; network infrastructure devices) to perform containment actions.

23. How well does your organization perform restoration activities to ensure the operational availability of its systems and services that may be affected by cybersecurity incidents?
    o   Incident Recovery Plan Execution (RC.RP)
    o   The organization's restoration activities may include using the results of a business impact analysis or system categorization to validate that essential services are restored in an appropriate order.

24. How well does your organization coordinate restoration activities with internal and/or external parties?
    o   Incident Recovery Communication (RC.CO)
    o   The organization must communicate their planned or ongoing restoration activities and their current status.
    o   Robust communication during the recovery process should be utilized and senior decision makers should be kept up to date during the recovery from a major incident.

STEP 5: DETERMINE YOUR TARGET SCORES

With a clear understanding of the questions and the zero to ten scoring system, your next crucial step is to establish specific targets within the range of five to eight for each of the six functions outlined in the NIST Cybersecurity Framework: Govern, Identify, Protect, Detect, Respond, and Recover.

Recognize that your organization is unique, and your cyber risk requirements may differ from others. Consequently, it is important to prioritize certain aspects of cybersecurity based on your organization's needs.

While a score of five represents the minimum acceptable level, and eight signifies full optimization, your organization has the flexibility to choose scores that it deems appropriate and reasonable in alignment with its perception of the threat landscape and risk tolerance. This customized approach ensures that your cyber risk management efforts align precisely with your organization's threat landscape and risk appetite.

When determining its target scores, the organization can choose one of several different approaches, including the minimum score approach, the strong castle approach, the first responder approach, the big city approach, or the world-class approach. Each of these approaches has its own advantages and disadvantages, too.

The **minimum score approach** sets out to achieve a minimum score across the board based on the belief that this is reasonable within the organization's industry, customer expectations, and organizational maturity. In a minimum score approach, each of the six functions is assigned a target score within the minimum acceptable range. This means that the Govern,

Identify, Protect, Detect, Respond, and Recover are each assigned a target profile score of five.

FUNCTION	TARGET
GOVERN (GV)	5.0
IDENTIFY (ID)	5.0
PROTECT (PR)	5.0
DETECT (DE)	5.0
RESPOND (RS)	5.0
RECOVER (RC)	5.0

1	2	3	4	5	6	7	8	9	10

Example of Minimum Score Approach

The **strong castle approach** selects a target score profile emphasizing the Protect function. In a strong castle approach, each functional area is assigned a value of 5 (the minimum acceptable score) except for the Protect function, which receives a higher target score of 7 or 8.

FUNCTION	TARGET
GOVERN (GV)	5.0
IDENTIFY (ID)	5.0
PROTECT (PR)	7.0
DETECT (DE)	5.0
RESPOND (RS)	5.0
RECOVER (RC)	5.0

1	2	3	4	5	6	7	8	9	10

Example of Strong Castle Approach

The idea behind the strong castle approach is to identify the most critical things to protect and then focus your time, effort, and resources on protecting those things. This strong castle approach and its associated target score profile is still the current practice of many organizations across various industries. When implemented well, this strategy should minimize the need for optimal cybersecurity in the other functions since you've already assumed that only minor incidents would occur within them.

We call this the strong castle approach because it is similar to a medieval castle or fortress. When it was built, these structures could withstand any attack currently available to its occupants' enemies: arrows, battering rams, swords, etc.

But as weaponry evolved and progressed, a castle or fortress could no longer withstand any attack. Could you imagine a medieval castle surviving a missile attack from a drone?

Well, it isn't likely, and that's why the strong-castle approach to cybersecurity has been declining in popularity in recent years. As cyber attackers have become more effective at compromising people rather than just the organization's technology, this strong castle approach is becoming less and less effective in our modern networks.

The **first responder approach** is an approach that sets the target score for the Respond function as a 7 or 8 while allowing relatively lower target scores in the other five functional areas. This approach focuses the organization's resources on building out a fast, high-quality response capability in order to mitigate the other five functional areas having lower target scores.

FUNCTION	TARGET
GOVERN (GV)	5.0
IDENTIFY (ID)	5.0
PROTECT (PR)	5.0
DETECT (DE)	5.0
RESPOND (RS)	7.0
RECOVER (RC)	5.0

| 1 | 2 | 3 | 4 | 5 | 6 | 7 | 8 | 9 | 10 |

Example of First Responder Approach

The **big city approach** is a modern and mature perspective to cybersecurity in which the organization's data network is viewed as a modern city as opposed to a medieval fortress. Like a big city, the organization prioritizes the Respond and Recovery functions instead of heavily focusing on the Govern, Identify, Protect, and Detect functions. In the big city approach, a zero-trust mentality is required where any user is treated as untrusted, and the organization is poised to respond and recover as soon as an incident occurs.

FUNCTION	TARGET
GOVERN (GV)	5.0
IDENTIFY (ID)	5.0
PROTECT (PR)	5.0
DETECT (DE)	5.0
RESPOND (RS)	7.0
RECOVER (RC)	7.0

1	2	3	4	5	6	7	8	9	10

Example of Big City Approach

The **world-class approach** is one in which every functional area is treated as equally important, and a target score of 8 is assigned to all six functional areas. Being world-class at cyber risk management is very expensive and difficult to achieve. Oddly, it's only practical for either very small organizations or a government operation, such as the National Security Agency (NSA), the Central Intelligence Agency (CIA), and others who can spend whatever is needed to achieve this level of cybersecurity.

FUNCTION	TARGET
GOVERN (GV)	8.0
IDENTIFY (ID)	8.0
PROTECT (PR)	8.0
DETECT (DE)	8.0
RESPOND (RS)	8.0
RECOVER (RC)	8.0

1	2	3	4	5	6	7	8	9	10

Example of World Class Approach

While the minimum score approach, the strong castle approach, the first responder approach, the big city approach, and the world-class approach are all valid options, your organization doesn't have to pick a single named approach to utilize. Instead, they can choose any target score for any functional area they determine is important to their specific circumstances.

For example, let's assume you are working with a software as a service (SaaS) business with around $10 million in annual revenue. In that case, this organization may be focused on its ability to maintain records of its sensitive digital assets, like its customer's names, mailing addresses, and credit card information. Additionally, this organization likely has its own source code and trade secrets that must be protected. In this scenario, you may want to assign a target profile score of 6 or 7 to the Protect function while assigning a minimum score of 5 to the other five functions.

Conversely, if you are working with a geographically dispersed $800 million annual revenue company, the big city approach, focusing on response and recovery, might be more appropriate.

Regardless of your specific reasoning for selecting a certain approach or the assigned target profile score, you should take the time to record it before you begin conducting the interviews with the cyber risk influencers within the target organization.

STEP 6: CONDUCT THE INTERVIEWS

It is now time to conduct interviews with our cyber risk influencers from the organization. It might be tempting to simply send the 24-item questionnaire we developed earlier to the organization's employees in a mass email, but that is not ideal.

In fact, the best way to conduct the interviews is by doing them in person, or by video call, with a select group on an individual level. The very process of asking these questions is a marvelous training opportunity for the interviewee because most people don't know the definition of good cyber risk management and have, therefore, never even considered measuring it. It's also best if you're not the one conducting the interviews either. Instead, having a well-respected senior employee or a neutral outsider conduct the interviews is better.

During the interview, give each interviewee a printed copy of your score key. That way, when you ask them each of the 24 questions out loud, they can easily choose the score that best reflects their experiences in your company. Moreover, having the scoring table on hand helps to keep your respondents' answers uniform, thus giving you more reliable data across all the people you will be interviewing.

As you prepare to conduct your interviews, it is important that you set proper expectations with the organization and your interviewees. In the ever-evolving business landscape, companies often face the challenge of adapting to changes in order to survive and thrive. When a large corporation becomes entrenched in a certain way of doing business, it may encounter a crisis that necessitates a transformative shift. This can involve strengthening cybersecurity measures or even discontinuing entire product lines.

An example of such a transformation occurred at IBM in the early 1990s, when they recognized the need to shift from hardware manufacturing to providing services. IBM changed its product offerings and redirected its focus to develop the necessary skillsets for the service industry. Companies that endure understand the importance of self-evolution, as failure to adapt can lead to stagnation and eventual bankruptcy, as exemplified by Kodak's struggle to transition to digital photography.

Undertaking significant organizational changes, such as modernizing its operations, culture, and customer-centric mindset, can be overwhelming for employees if not managed effectively. It is not uncommon for people to resist changes in cybersecurity practices. To overcome this resistance, it is crucial to emphasize the collaborative nature of the transformation and highlight that everyone has a role to play.

The interviews should foster a candid and respectful atmosphere while maintaining a good pace. Encourage employees to provide forthright responses, assuring them that their scores will remain confidential and that no specific comments or scores will be attributed to individuals. Valuable insights can emerge when individuals freely express their perspectives, including scores of zero or explanations supporting their assessments. Additionally, to prepare employees for the interview process, provide them with comprehensive information about what to expect.

By managing organizational changes actively, setting the right tone, and ensuring transparency and confidentiality, you can gather the best possible data and insights from your employees. This will enable you to navigate the cybersecurity transformation effectively and drive meaningful progress within your organization.

As you begin the interview, you should spend the initial ten minutes of each interview providing the interviewee with an overview of the process you are using. The interviewer should reiterate that this is a management improvement effort, emphasizing that it is not an external audit. Encourage the influencer to be candidly respectful in their responses and inquire if they have any questions regarding the zero to ten scoring scale. Take the time to provide them with the score key and a brief explanation of the scoring process and the purpose of the interview.

Throughout the interviews, you should maintain a brisk tempo and aim for a duration of less than an hour per interviewee. Avoid requesting any justifications for scores, but if interviewees voluntarily provide explanations, be sure to record them in the spreadsheet, as they can prove helpful during Phase Two of the CR-MAP process. These interviews should not be lengthy or arduous. Both you and the influencer desire an efficient process that respects their time and expertise while allowing them to leave the session with a positive impression of the process.

During the interviews, you may encounter respondents claiming to have limited cybersecurity knowledge. In such cases, acknowledge their perspective but emphasize that they possess unique insight within their own department and are more knowledgeable about its practices than anyone else, which is why they were chosen as one of the cyber risk influencers you have been asked to interview. Reiterate that their opinions are valuable, even if based on perceptions or partial information.

When it comes to organizational change management, it is crucial to understand that people are usually uncomfortable with proposed changes. Actively listening to their opinions from the outset goes a long way in securing their ongoing buy-in and fostering a collaborative atmosphere. When people feel heard, they also feel respected and will be more supportive of potential changes.

STEP 7: COMPILE AND AVERAGE THE SCORES

Once you've completed all the interviews and recorded the associated scores, your next step is to average them out and compare them to your target scores for each NIST CSF related function, category, and subcategory from your questionnaire.

In our engagements, we use a customized spreadsheet to record all of the answers and generate a radar diagram with the results. A sample of this is included below.

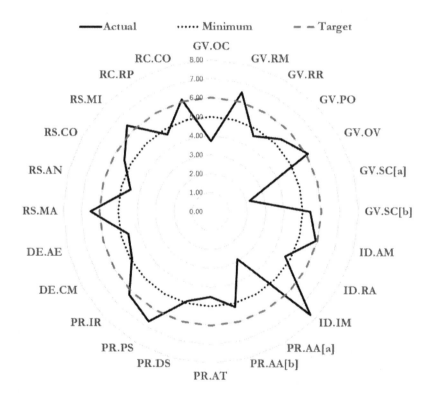

In this radar diagram, you will notice two circular rings. The outer dashed line represents the target score chosen by the organization. In this example, a target score of six was chosen for all six functions, but the organization could have chosen any number for each function individually that would meet the needs of their organization. The inner dotted line represents the minimum acceptable risk target score for this organization. Again, this organization has chosen to use a single score for all six functions, which in this case is a minimum acceptable risk target score of five.

Notice that there is a third line that appears solid and jagged in this radar diagram. This solid line represents the average response score received from the interviewees for each of the six functions and categories related to the 24 interview questions.

Based on the radar diagram, we can quickly identify this organization's top five cyber risks, including GV.SC[a] (average response 2.2), PR.AA[a] (average response of 2.9), GV.OC (average response of 3.7), RS.AN (average response of 4.5), and PR.AT (average response of 4.5). Notice all of these are well below the minimum acceptable risk target of 5 and our actual target of 6.

FUNCTION	ACTUAL	TARGET	GAP
GOVERN (GV)	4.8	6.0	1.2
IDENTIFY (ID)	6.1	6.0	-0.1
PROTECT (PR)	5.1	6.0	0.9
DETECT (DE)	4.8	6.0	1.3
RESPOND (RS)	5.7	6.0	0.3
RECOVER (RC)	5.4	6.0	0.6

| 1 | 2 | 3 | 4 | 5 | 6 | 7 | 8 | 9 | 10 |

Summary Scorecard

NIST CATEGORY (SORTED BY GAP SIZE)	ACTUAL	TARGET	GAP
GV.SC[a]	2.2	6.0	3.8
PR.AA[a]	2.9	6.0	3.1
GV-OC	3.7	6.0	2.3
RS.AN	4.5	6.0	1.5
PR.AT	4.5	6.0	1.5

| 1 | 2 | 3 | 4 | 5 | 6 | 7 | 8 | 9 | 10 |

Top 5 Cyber Risks

Notice the Top 5 Cyber Risks chart is sorted from the largest to smallest gap size (right-most column) and includes categories from the

Govern, Protect, and Respond functions. You may be wondering why the Detect, Identify, and Recover functions are not displayed in the chart above.

In our experience, most companies are relatively underinvested in some functions while being stronger in other functions. In this example, the organization's scores in the Detect and Recover functions had smaller gaps between their actual and targeted scores. This does not mean that the Detect and Recover functions are meeting or exceeding the target score, though. Instead, it simply means that these functions do not represent the most significant cyber security gaps within the organization based upon the interviewees' responses.

On the other hand, the Identify function did meet or exceed its target score. Since the Identify function was listed as achieving a score of 6.1 and its target was 6.0, this means that the Identify function is not a cybersecurity gap that needs to be addressed by the organization at this time.

Another thing to be on guard for in this seventh step of Phase One is what we refer to as junk data. Once the interviews are complete and the data has been collected, you will analyze that data to determine its true value. Sometimes, your interviewees may give you junk data, such as when interviewees simply respond with a score of eight or a similar high score without actually considering what that represents. This can be caused by a lack of knowledge on the organizational practices, overestimating the abilities of the organization's IT department, or simply wanting to get the interview over with as quickly as possible, but in each of these cases the result is junk data being collected during our interviews.

So, what should you do with this junk data?

The biggest mistake you could make is to simply delete the data, which is often people's first thought. Instead, the best course of action is to keep the data, but not include it in the calculations of the client's scores to prevent the junk data from skewing the organization's scores calculated during CR-MAP process.

Instead of using the data in calculations, think about why the data is unusable, and what that might say about the broader environment in the organization. Did all the executives give unrealistic scores for every question, or did all the operations personnel answer with an eight simply because they wanted to get their interviews finished quickly?

In either case, you need to uncover the root cause of this junk data and then incorporate it into the organization's CR-MAP results as a theme to be addressed in Phase Two or Phase Three of the process.

STEP 8: COMMUNICATE YOUR TOP FIVE CYBER RISKS

As we near the completion of the first phase of the Cyber Risk Management Action Plan, you may find that the results fall slightly short of your expectations. If so, you may seek additional details from the identified cyber risk influencers to understand why the actual scores deviated significantly from the organization's target scores. But, in general, most of your work in Phase One will now be complete, and it's time to analyze the data to uncover the narratives they might reveal.

From our experience, two prevalent themes often emerge: organizations tend to excel in the Protect function where they invest significant time and effort, and at the same time, most organizations are lagging behind in the Detect and Recover functions. Why is this the case?

Historically, cyber risks have been predominantly viewed as technological challenges, leading to a strong focus on the information technology departments' ability to maintain operational systems. As a result, they excel in detecting service outages, but often fall short in detecting potential breaches of sensitive corporate data. Additionally, organizations commonly struggle in the Recover function because they are unaccustomed to public scrutiny of their technological failures. Consequently, they lack the internal capacity to communicate technical breakdowns and data breaches effectively to the public. They often rely on marketing departments that may not possess the required expertise.

This brings us to a crucial point: it is possible that your target scores deviate significantly from expectations due to a perception influenced by fear-inducing headlines and narratives from vendors. It's important to note that the questionnaire we created does not directly address specific technological measures like firewalls. This was not an oversight in the questionnaire, but instead was a deliberate approach that stems from our perspective of treating cyber risks as business risks and using a top-down approach rather than a bottom-up approach starting from inside the IT organization. This top-down approach is also firmly rooted in the way NIST designed the Cybersecurity Framework to be used.

We recognize that while strong technological defenses, such as firewalls and anti-spam filters, are essential, relying solely on them will not guarantee success in today's cybersecurity landscape. Instead, our CR-MAP process guides you toward practicing reasonable cybersecurity while accounting for a broader range of factors that can help your organization achieve cyber resilience.

Unfortunately, sensationalized headlines of major cybersecurity breaches, like the widely documented Expeditors incident, and marketing tactics employed by vendors seeking to sell their products often provide an incomplete picture. However, with the scores obtained from your questionnaires, you possess the complete story specific to your company or organization, which truly matters in shaping your cyber risk management strategy.

SUMMARY

In this chapter, we delved into the first phase of the Cyber Risk Management Action Plan (CR-MAP) process, which focuses on discovering the top cyber risks faced by your organization. We discussed the importance of prioritizing risks effectively, given limited resources. We emphasized the need to widen the scope of cybersecurity considerations beyond technological defenses alone.

We explored the significance of setting proper expectations when initiating organizational changes, drawing insights from the experiences of companies like Expeditors, IBM, and Kodak. It became evident that actively managing the changes and ensuring employee buy-in is crucial for successful cybersecurity improvements.

There is also a strong need for effective communication during the interview process, emphasizing a candid and respectful approach to encourage honest responses. We emphasized the brisk tempo of the interviews and the importance of respecting the interviewees' time and expertise. Additionally, we addressed the importance of gathering as much information as possible, including providing resources to help interviewees understand the scoring and interview process.

We then discussed the importance of determining specific targets within the five-to-eight range for each function and activity in the NIST Cybersecurity Framework. This tailored approach ensures that your organization's cyber risk management efforts align with its unique needs and risk landscape.

Lastly, we explored the significance of effectively communicating your top five cyber risks. We identified common themes observed in organizations, such as excelling in the Protect function but struggling in the Detect and Recover functions. We highlighted the need to view cyber risks as business risks rather than purely technological challenges. We stressed the importance of relying on a comprehensive understanding of your organization's specific risks rather than being influenced solely by fear-mongering headlines or vendor narratives.

By following these steps and prioritizing cybersecurity considerations beyond technology, organizations can gain valuable insights, make informed decisions, and lay the foundation for effective cyber risk management in the subsequent phases.

CHAPTER THIRTEEN

PHASE TWO: CREATING A CR-MAP

In 2016, AsusTek, a computer hardware company, learned of some vulnerabilities in their routers. Instead of disclosing the vulnerability to the public, they hid it from their consumers. They continued selling the defective—and, frankly, dangerous—routers. As a result, widespread exploitation by hackers occurred, and the attackers were able to gain access to more than 12,900 connected storage devices.

AsusTek didn't notify their consumers or their retailers because they were concerned that if they did, their reputation would be diminished in the marketplace, which would negatively impact their sales. Their decision to prioritize reputation over consumer safety backfired, as disgruntled customers took control of the narrative, leading to greater long-term damage to the company's sales and reputation. As usual in these situations, someone found out about the vulnerability anyway, and rather than the company being in control of the story, the disgruntled consumers controlled the narrative in the press, and this hurt the company's sales even more in the long run than it would have if they just came clean and disclosed the issue.

Remember, if your organization discloses the vulnerability itself, you get to control the narrative. In fact, you not only control the narrative, but also build trust and enhance your brand image by taking ownership of the issue and communicating openly. For example, if AsusTek had made a public announcement saying, "We are very sorry, we have identified a problem with our routers, and here is our solution to remedy the problem," it could have actually been a brand-enhancing maneuver.

People love authenticity and vulnerability. That announcement might have resulted in a decrease in sales for a short time, but over the long haul, it would have enhanced their reputation, not diminished it. They would have become known as the straight talkers of cybersecurity. People would want to buy their products—not because they'd expect them to be flawless, but because they would know that when a flaw is discovered, they will be upfront about it and accountable. This is called **responsible disclosure**.

To illustrate the impact of responsible disclosure, we can draw a quick comparison between Home Depot and Target, both of which experienced significant data breaches over the years. In 2014, Home Depot suffered a data breach that resulted in the theft of 56 million stolen credit card records from their customer database. A year earlier, in 2013, Target suffered a similar data breach that caused 40 million stolen credit card records to be stolen.

Despite similarities in the breaches, Home Depot's prompt and transparent response garnered a more favorable public perception compared to Target, which delayed informing customers for a week. Home Depot's proactive approach mitigated financial consequences and maintained consumer trust, whereas Target faced greater criticism and financial repercussions.

After both data breaches were fully responded to and recovered from, the accountants tallied up the cost of each breach. Home Depot suffered $179 million from its data breach. On the other hand, Target ended up spending $292 million on its data breach, and its CEO of 35 years also got fired due to the incident. Responsible disclosure isn't just the right thing to do; it can actually save your organization a lot of money in the long run.

Practicing reasonable cybersecurity measures and promptly addressing data breaches are crucial components of a robust Cyber Risk Management Action Plan. While perfection may be unattainable, being forthright, recovering quickly, and maintaining open communication with customers can significantly minimize the damage caused by cyberattacks. By learning from these examples and adopting a transparent approach, organizations can navigate data breaches more effectively, safeguard their business, and preserve their reputation.

KEYS TO SUCCESS IN PHASE TWO

Before we create our Cyber Risk Management Action Plan, it is important that we look at some things that are the keys to successfully being able to develop your CR-MAP in Phase Two of the process. This includes prioritization; roles and responsibilities; understanding the scale; keeping it simple; and controlling the rate of change.

First, prioritization. Prioritization is a critical aspect of managing cyber risks as organizations face numerous risks with limited resources. Making tough choices and prioritizing both the risks and mitigations is essential. Although prioritization doesn't guarantee complete protection, it provides a reasonable approach to address the challenge. Trusting your prioritization and implementing the plan may be challenging, as others may question your choices, but it's important to listen to feedback while staying committed to your chosen path, knowing that adjustments can be made as needed during the third phase of the CR-MAP process if required.

Second, roles and responsibilities. Assigning roles to everyone in your organization and distributing the cybersecurity responsibilities across those roles is crucial. You should assess each job description to determine feasible additions to ensure that cybersecurity work is effectively distributed across the organizational staff. When assigning cybersecurity responsibilities, ensure that individuals are set up for success by providing the necessary resources and ensuring those roles have any required additional tools and support. For example, if customer service employees need to use unique passwords, systematize the process by including it in their job description and equipping them with a high-quality password manager with centralized reporting. Then, continue to monitor their usage of the password manager through a single console or program that allows you to assess compliance and address any gaps in cyber hygiene or job performance.

Third, understanding scale. When implementing changes as part of your Cyber Risk Management Action Plan, individual adjustments are typically simple and straightforward. For instance, adding a line to each job description can be delegated to your management team. However, as you strive to enhance the 24 questions and make larger-scale improvements, implementing changes across the entire company can become more complex, costly, and time-consuming. While a single person on the customer service team can quickly adopt a password manager for their accounts, scaling these changes becomes increasingly challenging as the size of the organization grows. Scaling each element of your action plan becomes more difficult with a larger organization, requiring careful consideration of complexity, costs, and time investments, so keep this in mind as you build for scale across the organization.

Fourth, keep it simple. The complexity of cybersecurity change often overwhelms organizations, leading to unclear priorities and benefits. However, the methodology presented in this book offers a solution by utilizing prioritization to provide the most effective approach to implementing cybersecurity changes in a highly focused and simplified manner. By maintaining a laser-like focus, the complexity is minimized to achieve the organization's desired target scores. This is where the data collected back in the first phase becomes crucial. If your staff is hesitant to implement additional security measures, you can present them with the numbers they provided, highlighting the need for action based on their own input.

Fifth, controlling the rate of change. When implementing your plan, it is important to avoid overwhelming your workforce with too many changes at once. Just like with other improvement programs, such as transitioning accounting software or email systems, the changes required for your cyber risk mitigations should be carefully sequenced at a manageable rate alongside other ongoing changes. Determining the appropriate pace of implementation is a decision specific to your organization.

The good news is that not all mitigations will directly involve active changes from your staff. As you will discover, some mitigations will be transparent to them, allowing for their seamless integration alongside the more significant changes. This can help increase the rate of change across the organization without causing a backlash against the Cyber Risk Management Action Plan and its recommended risk mitigations that need to be implemented.

DEVELOPING AN ACTION PLAN

Now that you have identified and prioritized your top five cybersecurity risks, it's time to develop your Cyber Risk Management Action Plan. This crucial second phase will span thirty days, during which you will take significant strides toward enhancing your organization's cybersecurity posture.

It is important to acknowledge that actively managing every cyber risk may not be feasible or necessary. Simply going through the questionnaires and identifying your risks already puts you ahead of many other organizations and competitors who have yet to take similar steps.

However, to safeguard yourself in case these risks materialize differently than anticipated, it is recommended to work under attorney-client privilege. This affords some protection to your organization if a cyber risk you didn't prioritize becomes the source of a data breach, guarding against potential accusations of negligence both legally and in the court of public opinion.

Another important factor to consider is your contractual limitations regarding and obligations to third-party vendors, contractors, and suppliers. Many organizations, such as credit card processors, rely on vendors to deliver products and services. However, if a vendor experiences a significant cyber failure, such as a prolonged outage or a breach of sensitive information, it can severely affect your reputation and customer trust.

To mitigate potential risks, you have to communicate your expectations to vendors and customers regarding their responsibilities in protecting against and addressing cyber failures. This can be achieved through well-crafted contract language that establishes indemnification clauses, creating a contractual firewall for your organization. Indemnify here means compensating someone for harm or loss.

It is crucial to note that while we can provide guidance as cybersecurity practitioners, it is essential that you consult with a licensed, qualified attorney to obtain specific legal advice tailored to your organization's unique circumstances. Based upon decades of experience, we strongly recommend that you do not rely on forms found online or ones created by artificial intelligence tools as they are often incomplete, overbroad, include inaccurate statements of law, and are not tailored to your organization's specific needs. In the end, using these types of shortcuts will often cost you more in the end in terms of time, finances, and energy.

When engaging with vendors, ensure that your master services agreement or similar written contract clearly outlines shared responsibility for data security. This written agreement should include an indemnification provision specifying the financial responsibility in the event of a cybersecurity failure and defining limits to that responsibility. If the vendor is accountable for a failure, contractually require that they are legally responsible for covering costs related to liability, legal defense, and crisis management, encompassing first-party and third-party expenses.

In terms of customer expectations, it is prudent to limit your liability in the event of a cybersecurity failure. To do this, you should state that your service offerings are provided as is, and your liability is restricted to the actual amount customers have paid for the services.

Remember, consulting with a qualified, licensed lawyer will ensure that your contractual agreements and limitations are legally sound, enforceable, and adequately protect your organization's interests.

Now, to develop your Cyber Risk Management Action Plan, you will follow five steps inside of Phase Two of the CR-MAP process:

1. Close the identified gaps
2. Conduct a business value analysis
3. Create a dashboard and roadmap
4. Conduct internal marketing
5. Conduct external marketing

STEP 1: CLOSE THE IDENTIFIED GAPS

Begin by focusing on your top-ranked risk and examining the disparity between your actual scores and target scores. Take a moment to reflect on the actions required to bridge this gap. It's important to note that achieving a target score of 5 doesn't necessitate investing in top-of-the-line cybersecurity capabilities.

Instead, the key is to determine what is reasonable and appropriate for your organization. This forms the fundamental question for resource allocation in managing your cyber risks: What specific measures does my company need to undertake in order to reach our target?

Consider the following Top 5 Cyber Risks chart:

NIST SUBCATEGORY (SORTED BY GAP SIZE)	ACTUAL	TARGET	GAP
GV.SC[a]	2.2	6.0	3.8
PR.AA[a]	2.9	6.0	3.1
GV-OC	3.7	6.0	2.3
RS.AN	4.5	6.0	1.5
PR.AT	4.5	6.0	1.5

| 1 | 2 | 3 | 4 | 5 | 6 | 7 | 8 | 9 | 10 |

Top 5 Cyber Risks

Notice that the top cyber risk identified was GV.SC[a]. This identified subcategory is found under the Govern function and the Cybersecurity Supply Chain Risk Management subcategory. This risk was identified during the interview process when the respondents were asked,

"How well does your organization identify, establish, manage, monitor, and improve cybersecurity supply chain risk management processes?"

The average response from our interviewees was 2.2, while the target score was 6, and the minimum acceptable score was 5. This leaves the organization with a gap of 3.8 (6 target score − 2.2 average response).

To close this identified gap, we should first look at the NIST Cybersecurity Framework to determine the requirements to meet its outcome. When working with the NIST Cybersecurity Framework requirements, we often reword them into questions or testable statements to make them easier to use during our engagements, as we did with GV-SC[a].

To achieve your target score of six, it is essential to thoroughly analyze each question and determine how to ensure that each requirement is met at a level of six. Once you have made the necessary improvements, the goal is to confidently state that your organization consistently fulfills these requirements with minimal flaws. Developing specific steps for improvement demands specialized expertise, so it's important not to expect to tackle it alone. Seeking assistance is crucial, especially considering the need for customization to align with your unique security systems and business model.

While we cannot provide a one-size-fits-all solution due to these considerations, here are some controls and activities you could implement to aid in improving the organization's score for the GV.SC[a] category:

- Develop a Supply Chain Risk Management Policy: Create a comprehensive policy that outlines how the organization will identify, assess, manage, monitor, and improve cybersecurity risks within its supply chain. This policy should define roles, responsibilities, and procedures for conducting risk assessments on suppliers and integrating cybersecurity risk management into the procurement process. Regularly review and update this policy to reflect changes in the threat landscape and business practices.

- Implement Supplier Security Assessments: Conduct thorough security assessments of all current and potential suppliers to evaluate their cybersecurity practices and controls. Use standardized assessment questionnaires and, where possible, on-site audits to gain a detailed understanding of suppliers' cybersecurity postures. Assessments should be conducted regularly and whenever significant changes occur within the supplier's organization or to the services they provide.

- Establish Continuous Monitoring of Suppliers: Implement tools and procedures for the continuous monitoring of key suppliers' cybersecurity practices and incident reporting. This can include the use of shared threat intelligence platforms, monitoring of public data breaches involving suppliers, and requiring suppliers to report any security incidents that may impact the organization.

- Supplier Performance Metrics and Contractual Obligations: Integrate cybersecurity requirements and performance metrics into contracts with suppliers. This should include obligations for suppliers to adhere to specific cybersecurity standards, timely reporting of security incidents, and regular provision of evidence demonstrating compliance with agreed-upon cybersecurity practices. Regularly review suppliers' performance against these metrics to ensure compliance and address any gaps.

- Supply Chain Risk Awareness and Training: Develop and deliver a comprehensive training program for all employees involved in procurement, supplier management, and IT security roles to raise awareness of supply chain risks and the importance of cybersecurity risk management in the supply chain. The training should cover the organization's policies and procedures for managing supply chain risks, how to conduct supplier assessments, and how to respond to incidents involving suppliers.

This process should be completed for each of your identified top cyber risks to ensure each has been thoroughly considered, and appropriate controls and mitigations are recommended for inclusion in the Cyber Risk Management Action Plan currently under development.

IN-HOUSE VERSUS OUTSOURCED TASKS

These days, cybersecurity experts are in high demand, and research shows that the trend will continue for years to come. Unfortunately, the sources of cybersecurity talent have been unable to keep pace with the dramatic rise in cybercrime.

Particularly rare are cybersecurity professionals who understand and practice the major point made within this book: cyber risks are business risks just as serious and worthy of the executive leader's attention as risks to sales, order fulfillment, and accounts receivable. This means hiring the talented cybersecurity people you want on your team may be difficult. Even if you can find them, they will be expensive. Also, odds are they will receive frequent, unsolicited job offers from organizations willing to pay more than you are, in which case you will lose them from your team.

For example, in the Seattle area, many large, growing employers in the technology industry are paying very high compensation for experienced cybersecurity people. That makes it difficult for smaller, non-technical organizations to hire and retain the cybersecurity staff they need. Some of the struggling organizations are opening smaller offices in other cities or countries to find the talent they need. Or, they switch to outsourcing.

This means you need to be very smart about where to get the talent you need to execute your Cyber Risk Management Action Plan. To help mitigate this talent gap, we recommend carefully considering which work you want to keep in-house versus which work will be outsourced.

The general guidance for this is simple: keep the work that is core to your business in the hands of your trusted insiders and employees.

In contrast, there are lots of tasks that good candidates outsource to another company. To help determine which tasks should be worked on by which set of experts, we recommend splitting all of your tasks into one of three categories: core tasks, strategic outsourcing tasks, or commodity outsourced tasks.

Core tasks are ones in which your employees perform tasks that help the business take smart cyber risks, deliver higher quality cybersecurity decisions than outsiders would be able to, and establish and maintain critical business relationships with people across the company so the cybersecurity agenda gets the attention it deserves.

Strategic outsourcing tasks are where the employees are directly assisted by outside experts who do the majority (60–80%) of the detailed work. For example, if your organization hired a consultant to lead the development of your first Cyber Risk Management Action Plan, that would be classified as a strategic outsourcing task.

Commodity outsourced tasks are ones in which the outsiders do all (100%) the work under the direct oversight of your employee. For example, your organization may hire an outside firm to conduct your quarterly PCI-DSS assessment vulnerability scans. This is a commodity outsourcing task because any certified provider could do it equally well and with limited oversight from your employees. Therefore, we can shop around for the best price or overall bundle of services to complete this task since it is heavily commoditized.

Let's consider how these three categories could be used to separate some notional tasks.

First, we have the core tasks. Maybe we need to bring someone on to work as a project manager, preferably with IT security knowledge and experience, which can be learned. Since it is a tight labor market, we may opt to provide an internal opportunity to an existing employee first when trying to fill this position. This project manager will be focused on several core tasks, including committee work; chairing the cyber risk committee; participating on the change control committee; participating on the disaster recovery committee; developing and maintaining standards and procedures; preparing for and supporting annual external audits; supporting the annual cyber insurance renewal; conducting information assessments; and many others.

In addition to those tasks, we can consider several strategic outsourcing tasks, including developing and maintaining organizational policies; annual company cyber risk assessments; high-risk security assessments; annual firewall effectiveness assessments; technical vulnerability assessments; and many others.

Lastly, we have the commodity outsource tasks, which include conducting network intrusion detection activities; performing education and awareness training; administering anti-phishing training programs; performing periodic control reviews; and more.

COST OF IMPLEMENTATION

Whether you decide to keep cybersecurity services in-house or outsource them, it is important to calculate the implementation costs. In our cyber risk management planning, we utilize the three-year total cost of ownership (**3TCO**) to estimate the overall costs of each mitigation. This allows for a better comparison of costs across different mitigations, considering one-time costs, ongoing expenses, and varying cost structures.

The formula to calculate the 3TCO is:

(Implementation Cost) + (Annual Operating Cost x 3)

Implementation cost refers to the total expenses associated with acquiring and deploying a system, service, or solution, including acquisition costs and labor expenses for implementation. If you need to calculate the implementation cost, you can do this using the following formula:

(Acquisition Cost) + (Implementation hours x Labor Cost)

When the specific resource responsible for the work is unknown, a blended labor cost of $125 USD per hour is typically used, combining internal and external labor rates. But, this figure does change over time and based on the current labor rates applicable to your market. Once the responsible resource is identified, the cost estimate can be updated accordingly, utilizing the appropriate labor costs.

The annual operating cost is another important number to consider. **Annual operating cost** refers to the total expenses incurred on an annual basis to maintain and operate a particular system, service, or solution, including renewal costs and ongoing labor expenses. The annual operating cost is calculated using the formula:

(Annual Renewal Cost) + (Operating Hours × Labor Cost)

To illustrate this approach, let's consider an example where you need to implement a new security event log management system to enhance your ability to detect cybersecurity incidents quickly.

Let's consider an example where operating a given system or control would take two hours per week. This would equate to 104 hours per year at a cost of $13,000 ($13,000 = 104 hours x $125/hour). The system also charges an annual licensing fee of $7,200 per year to operate. This means the total operating cost per year is $20,200 ($20,200 = $13,000 labor/year + $7,200 software/year).

But, this is only the cost per year for a single year, not 3 years. Therefore, the total operating costs would be 3 x $20,200, a total of $60,600 to implement this control.

Unfortunately, we forgot about the upfront implementation cost, so we must also add that. The implementation cost is $36,000 to acquire the system, 80 hours to implement it, and the implementation labor rate is $125/hour. This means we have a total implementation cost, or one-time fee, of $46,000 ($36,000 + (80 hours x $125/hour)).

So, the total cost of ownership over the first 3-year cycle (3TCO) would be $106,600. This is $46,000 for the implementation cost plus another $60,600 for the operating cost over the 3 years.

Let's consider another example. This time, we are working with an organization that needs to implement a new crisis communication plan to help the executive management team retain control of the organization when its primary communication systems are down during a ransomware or another type of cyber attack.

For this organization, we are essentially rewriting policies and determining which systems to utilize as opposed to buying and implementing a new system. This means we have $0 in software licensing or system acquisition costs, but we will have 160 hours of labor required to build out the new crisis communication plan at the cost of $125/hour. This gives us a total implementation cost of $20,000 (160 hours x $125/hour).

The organization will spend about 40 hours per year to operate and maintain this new plan at a cost of $125/hour, generating $5,000 per year in operating costs.

To calculate the 3TCO, we will add the implementation cost ($20,000) to three times the annual operating cost (3 x $5,000), giving us a total of $35,000 to acquire/develop and operate this new crisis communication plan.

Using these formulas, you should be able to calculate the cost of any new system, control, or mitigation that you may want to recommend as part of your new Cyber Risk Management Action Plan using a three-year total cost of ownership (3TCO) figure for easy comparison against other potential solutions.

STEP 2: BUSINESS VALUE ANALYSIS

Your mitigations can create value for your company in four dimensions: financial returns, technical risk mitigation, legal risk mitigation, or increased reliability of operations.

The purpose of doing a business value analysis is to take what could otherwise be an obscure set of decisions related to cybersecurity and allow you to communicate the benefits of those actions to key decision-makers and later to everyone else in more business-friendly terms.

The Business Value Model contains trust at its core and divides the potential business value into four quadrants. Although the diagram shows this as four equal quadrants, your organization can increase or decrease the importance it places on each section based upon your organizational needs.

Business Value Model

The Financial Returns quadrant focuses on cost savings, competitive differentiation, increased productivity, decision enhancement, and brand enhancement.

The Technical Risk Mitigation quadrant focuses on data confidentiality, trustworthiness, authorization, and business continuity.

The Legal Risk Mitigation quadrant focuses on due diligence, increased accountability, external compliance, internal compliance, and improved awareness.

The Increased Reliability of Operations quadrant focuses on increased availability, preservation of data integrity, disaster recovery, and preservation of current capabilities.

Let's consider the example of a password manager being recommended for implementation across the customer service team of an organization.

The customer service director doesn't know much about technology, cyber risks, or password protection. After all, that isn't their job. Instead, their job is to serve customers.

If you go to the director and say, "We are rolling out a new cybersecurity process for your department that will require a new piece of software be installed on all of your customer service agent's workstations," you will likely get a bad reaction from them.

In the past, we've heard responses like, "Why are you making my people do this?" or "They're already busy enough; how is this plan of yours going to help us serve our customers better?" or even "I don't see the value in that."

In short, they will protect their team from disruptive changes because they need to keep their team's productivity levels high. To be persuasive in that situation, you must use language that makes sense to them. By using the business value analysis model to explain your proposed changes, you can make them understand how this change will make their entire department better and, hopefully, more productive.

Remember, if you cannot explain the purpose behind your cybersecurity change in a way that gets buy-in from the business side of your organization, your changes are sure to fail. Soft skills are extremely important during this part of the CR-MAP process.

Another way the business value model can be useful is when you need to justify spending money on cybersecurity measures. For example, a password manager for the entire customer service department represents a $25,000 annual expense. Since that money must come from somewhere else in the organization's budget, you will compete with other proposals for how your company should spend that same $25,000 as money is a finite resource.

Someone from the marketing department might propose allocating that $25,000 to run a marketing campaign that would increase your top-line revenue by 2 percent. Someone from sales might say they could hire three interns for the summer with that money and increase their number of outbound calls by 5 percent. At the end of the day, you are competing with all of those proposals. Hence, you need a simple, straightforward example of how your cybersecurity measures will bring value to the business and is a valid way to spend the organization's valuable and limited resources.

Let's analyze the business value of implementing the crisis communication plan mentioned earlier in this section.

The crisis communication plan has its major business benefit in the technical risk mitigation quadrant. We can identify several applicable value factors for this proposal, including data confidentiality, trustworthiness, and business continuity.

Implementing the crisis communication plan will reduce the risk of unauthorized disclosure, avoid breach notification costs, and mitigate the risk of regulatory action, leading to increased data confidentiality. The crisis communication plan will also increase trustworthiness in the organization by enhancing confidence in the overall security of systems and processes. The plan will also improve the business continuity capabilities of the organization by allowing it to recover and sustain critical business functions and customer processes after a disaster.

While other benefits, such as decision enhancement, may exist, technical risk mitigation aligns most strongly with the objectives. Now that we have identified these factors and any additional ones we might want to add, we build support for implementing the crisis communication plan to minimize the gap between our target score and our currently assessed score within the organization.

STEP 3: DASHBOARD AND ROADMAP

It is time to consolidate all the organization's cyber risk mitigations into a single concise dashboard and overarching roadmap. This will help us to incorporate all the various forms of analysis we have already performed, including the three-year total cost of ownership (3TCO), the primary business benefits, and any secondary (or optional) business benefits you have identified. This dashboard will serve as a prioritized list of actions, arranged based on the magnitude of the gap each mitigation will address.

To enhance your execution efficiency, you may also consider developing an implementation roadmap alongside the dashboard. While the specifics of creating roadmaps can vary among organizations, it's important to find a method that suits your organization's specific needs.

Having worked with numerous organizations over the years, we have seen numerous roadmap templates and styles, from traditional Gantt charts, Kanban boards, Scrum product backlogs, and other Agile tools. Additionally, you can use any template or format your organization prefers. You should always consult your project management team for additional guidance on the roadmap format.

If you're unsure or just looking to create your own roadmap, we recommend a quick web search for the term "Gantt chart Excel" to find several different templates and tutorials to help get your roadmap development started.

Priority	Mitigation Name	Business Value	Estimated Implementation Cost	Estimated Maintenance Costs	Three-year TCO
1	Implement cybersecurity training company-wide	Risk, Legal	$30,440	$38,440	$145,760
2	Activate auto-encrypt of USB storage company-wide	Risk, Reliability	$2,500	$0	$2,500
3	Revise security requirements in contracts	Risk, Legal	$6,400	$0	$6,400
4	Implement password manager company-wide	Risk, Reliability	$27,940	$38,440	$143,260
5	Improve governance of cybersecurity	Risk, Legal	$51,800	$49,700	$101,500
6	Implement Vulnerability Scanning	Risk, Reliability	$10,000	$18,000	$64,000
7	Implement two-factor authentication company-wide	Risk, Legal	$5,000	$0	$5,000

Implementation Roadmap

As you craft your implementation roadmap, you may be tempted to alter the order of the mitigations. If you decide to do so, ensure you have a reasonable rationale for making that change, and you should also ensure that your rationale is documented for future reference. Remember that you may need to explain your decisions to important stakeholders such as major customers, investors, state regulators, or even in a legal setting someday in the future, so having your rationale documented will come in handy.

One valid justification for a change could be managing the change pace for the organization's staff. In the provided table, notice that mitigations #3 (contracts), #5 (governance), and #6 (scanning) have minimal impact on personnel. Adjusting their implementation sequence might be prudent if your employees require a respite from the influx of new cybersecurity practices being introduced by your recommendations.

STEP 4: INTERNAL MARKETING

Once you have identified the gaps that need to be addressed within your organization and assigned roles for everyone in your Cyber Risk Management Action Plan, securing buy-in from the organization's key stakeholders is crucial. This is where internal marketing plays a critical role in ensuring the success of your initiatives.

To effectively communicate and engage your company, consider addressing the following areas and sharing detailed information about them through various communication channels while leveraging established platforms such as newsletters and weekly team meetings. Collaborating with your marketing team can be advantageous as they possess valuable tools and techniques to support your efforts.

First, clarify the goals and objectives of implementing the next mitigation. It is important to clearly articulate the intended outcomes and the purpose behind the chosen approach because this helps employees understand the significance of the changes and align their efforts accordingly.

Next, you must establish how you will measure the success of the implemented mitigation. Define key performance indicators (KPIs) or metrics that will help track progress and determine if the desired outcomes are being achieved. These metrics will be used to create transparency and provide a clear benchmark for evaluating the effectiveness of the initiatives.

Then, you need to outline the timeline for implementing the proposed changes. Once outlined, you should clearly communicate the expected milestones and deadlines, allowing individuals to anticipate and plan accordingly. A well-defined timeline ensures the implementation stays on track and helps manage expectations throughout the process.

Next, you will need to identify the individuals who will take the lead in driving these changes. Designate responsible leaders or champions who will guide the implementation process, as well as those who will ensure that accountability and coordination amongst the teams occur. This clarity of roles and responsibilities encourages active participation and helps facilitate a smooth execution.

Then, you must assess whether new hardware or software will be required to support your new action plan. Determine if any technological investments or upgrades are necessary and communicate the rationale behind these decisions. This ensures that the required resources are available to implement the mitigation strategies effectively.

Next, you should consider your staff's training and skill development needs. Evaluate whether additional training programs or workshops are necessary to equip employees with the knowledge and skills needed to adapt to the changes. Communicate the availability of these learning opportunities to support employee growth and facilitate a smooth transition.

Lastly, the organization will need to address the cost implications of the mitigation efforts, both in terms of financial investment and time commitment. The associated costs must be clearly communicated, including any required budgetary considerations or resource allocations. This transparency enables stakeholders to understand the investment involved and appreciate the initiatives' value.

By effectively addressing these questions and utilizing diverse communication channels, you can engage your company, foster understanding, and gain the necessary support to successfully implement your cyber risk management initiatives.

STEP 5: EXTERNAL MARKETING

If you find yourself in a situation where you may need to explain your cybersecurity work to important stakeholders, such as your biggest customer, an investor, a state regulator, or even a judge and jury, it's essential to take additional steps to ensure your story is clear and comprehensive. Let us guide you through the process of preparing a one-page scorecard that can be used to efficiently present your entire cybersecurity narrative.

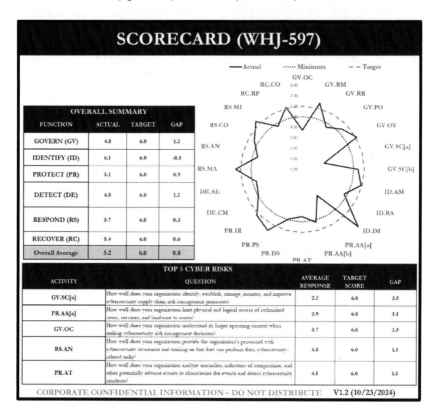

To create the Scorecard, you simply need to include: (1) the Overall Summary chart; (2) the Top Five Cyber Risks chart; and (2) the radar diagram. Collectively these charts and diagram visually represent the organization's cybersecurity posture. The exact layout and content can always be customized to meet the needs of your organization and your specific audience's needs. When constructing the scorecard, there are some important tips to consider.

First, it is recommended that you exclude the organization's name and logo from the Scorecard. This may seem counterintuitive, but this practice will help to prevent any temporary weaknesses or vulnerabilities disclosed in the Scorecard from falling into the wrong hands. Usually, these scorecards are shared both internally within the organization as well as externally with the organization's partners. Therefore, omitting the organization's name from the Scorecard is considered a good idea and a best practice. Additionally, it is crucial to strictly limit access to the data used in creating the Scorecard, by implementing need-to-know practices when working with this type of sensitive information.

If possible, obtain signed nondisclosure agreements (NDAs), drafted by a licensed attorney, from anyone who may view the Scorecard. This legal document provides your organization with legal recourse should the confidential information be improperly disclosed. You should also include a label on the Scorecard that indicates that the information contained therein is confidential. If your organization lacks specific labeling guidelines, a label such as "Confidential Information – Do Not Distribute" may be utilized.

It is essential to add the publication date to the footer to track the progress and version of the scorecard. This allows for clear identification and reflects any subsequent iterations or updates made.

Furthermore, as a security measure, random license plate codes like WHJ-597 can be assigned to the Scorecard. This anonymizes customers' cyber risk records and helps to safeguard their identity in the event of a breach or data loss.

By following these guidelines and creating a clear and concise Scorecard, you will be well-prepared to effectively communicate your cybersecurity efforts to key stakeholders while ensuring the confidentiality of the organization's sensitive information.

SUMMARY

Having identified and prioritized your company's top cyber risks, as well as developing a comprehensive plan to manage those risks, you have taken crucial steps to protect your organization and its stakeholders. This includes gaining buy-in from decision-makers and employees and determining the necessary budget for implementing the plan. Without these proactive changes, your company and customers risk significant consequences, such as the loss of valuable information, resources, damage to reputation, and even potential physical harm.

Many companies, like AsusTek, have made mistakes in the past, often failing to communicate their cybersecurity failures publicly. However, with the plan you have developed during the second phase of the Cyber Risk Management Action Plan process, not only can you significantly reduce the risk of cybersecurity breaches, but you can also leverage any breaches that do occur as an opportunity to build trust with your customers. You can demonstrate your commitment to cybersecurity and strengthen customer confidence by handling incidents transparently and effectively.

It is important to remember that implementing your plan and improving cyber hygiene is an ongoing process. Merely having a well-designed plan is not enough. You must continue to execute and maintain your Cyber Risk Management Action Plan to ensure long-term effectiveness. Cybersecurity is not simply about purchasing the right tools and technologies; it is about consistently implementing best practices and taking proactive measures to mitigate risks.

By staying vigilant and actively practicing cybersecurity measures, you can protect your organization, maintain customer trust, and minimize the potential impact of cyber threats.

CHAPTER FOURTEEN

PHASE THREE: MAINTENANCE AND UPDATES

In early 2017, an unprecedented cyberattack targeted Equifax, one of the three large credit bureaus in the United States. This attack highlighted their abysmal cyber hygiene practices at the time. The attackers, whose identities remain unknown, demonstrated a high level of sophistication given the scale and precision of the breach. They meticulously navigated Equifax's network, identifying valuable credit information, and subsequently exfiltrated gigabytes of data without detection. Regrettably, this breach compromised the credit files of countless working Americans, raising concerns about the long-term repercussions and the motivation behind the attack.

The nature of this breach suggests that it was not an opportunistic act or the work of a mere political activist. Instead, it appears to have been a deliberately calculated and silent operation to inflict maximum damage. The stolen data's uncertain fate hints at a larger, potentially more insidious motive. One possibility is that the data is being amassed for future exploitation. If a foreign government sought to undermine the United States, as witnessed during previous presidential election interference campaigns, the compromised credit information could be weaponized to compromise the entire credit-granting system and could undermine the American public's faith in the nation's banking and credit systems.

The consequences of such an influence attack on our economy would be dire. Imagine the repercussions if a significant number of credit files were tampered with or filled with fraudulent entries. Every day individuals seeking to purchase homes or cars would face insurmountable obstacles. Although the derogatory marks in their credit files would be unfounded, the economic gears of the country could grind to a halt, stifling growth and causing widespread disruption.

The Equifax incident carries with it immense implications. The company's response to the data breach was marred by numerous missteps, some of which were tragically ironic. In a misguided attempt to direct people to their site for information, Equifax inadvertently directed its Twitter followers to a phishing site by mistake. However, the most significant error was their failure to patch their Internet-facing web servers, exposing a shocking lack of cyber hygiene. The attackers did not need to resort to phishing tactics or zero-day vulnerabilities, but instead, they swiftly breached Equifax's network and operated undetected for a staggering one hundred days by exploiting a well-known and documented vulnerability in the Apache Struts framework.

Equifax, an organization entrusted with safeguarding vast amounts of sensitive and personally identifiable information about more than 140 million Americans, profoundly failed in its duty to protect consumers. The repercussions of this data breach reverberated throughout our community, impacting more than just a single company or individual. While your organization may not handle the same magnitude of sensitive data as Equifax did, you are still responsible for protecting your digital assets to the best of your abilities for the sake of yourself, your customers, and the broader community.

Achieving this level of protection necessitates more than a one-time implementation of a Cyber Risk Management Action Plan. It requires ongoing maintenance and updates to adapt to an ever-evolving threat landscape. The goal is to surpass the minimum standards and demonstrate a commitment to robust cybersecurity practices. By doing so, we can strive to safeguard our digital landscape and ensure a more secure future for our organization and its key stakeholders.

As we enter Phase Three of the Cyber Risk Management Action Plan (CR-MAP) process, we must note that these steps are not performed as strictly linearly as in previous phases. Instead, these steps are often conducted at the same time or using an iterative nature. The entirety of this third phase takes about ten months to complete, which will complete the full twelve-month CR-MAP process cycle.

Regularly conducting check-ins and reviews with your team is vital to sustaining your Cyber Risk Management Action Plan. These sessions serve multiple purposes, including assessing your progress toward achieving robust cybersecurity, acknowledging accomplishments, and updating your scorecard. It's crucial to remember that the scores obtained in Phase One of the CR-MAP process are not fixed or static. Instead, they should evolve over time for each item on the questionnaire and update them based on your implemented recommendations being fielded within the organization.

STEP 1: CONTINUALLY UPDATE YOUR SCORECARDS

While implementing change is important, documenting and tracking the progress is equally crucial. Therefore, it is advisable to regularly track your successes and create new scorecards every 90 days. Having up-to-date cybersecurity scorecards can serve as invaluable evidence in challenging conversations with regulators or customers, demonstrating your commitment to practicing reasonable cybersecurity.

Updating your scores not only provides data, but also tells a compelling story. For instance, when engaging with a regulator, you can showcase the progress made over time by stating, "A year ago, our average score for detecting cybersecurity incidents was 3.6 during our initial cyber risk assessment; however, through sustained efforts over five quarters, we have significantly improved our score to 5.1."

This type of narrative allows you to articulate the specific measures taken to enhance your cybersecurity posture, providing a data-driven story rather than relying on executives' unsubstantiated general statements when discussing their organization's cybersecurity posture.

STEP 2: SCHEDULE MONTHLY CHECK-INS

In addition to updating your cyber risk scores, it is essential to establish a pre-scheduled series of monthly cybersecurity check-ins for the entire year. These check-ins, which typically take about an hour, should be consistently held on the same day and time each month to maintain a sense of regularity and consistency.

When scheduling these meetings, being selective about the attendees is important. Focus on including key individuals who can significantly impact the success of your cybersecurity efforts. These meetings do not need to be large-scale gatherings; they should be kept small and tactical. The primary purpose is to monitor progress and identify any obstacles hindering your advancement. Having too many participants can hinder the efficiency of these meetings and the entire process.

Since you are already aware of the cybersecurity initiatives you are actively working on, these check-ins serve as an opportunity to update each other on the progress made in executing different parts of the Cyber Risk Management Action Plan. These meetings also serve as a chance to review the plan and ensure that you are aligned with the organization's objectives while providing an avenue for making necessary adjustments to stay on track.

If you find yourself off track during the monthly check-ins, the solution might be relatively simple. It could be a matter of lacking the necessary skill set to effectively mitigate a specific cyber risk. In such cases, it becomes a human resources challenge that can be addressed by hiring or contracting someone with the required expertise to help resolve the issue.

Similarly, if your company is undergoing significant corporate changes that consume a substantial amount of energy and resources, it may become difficult to handle multiple changes simultaneously. In such situations, it might be necessary to temporarily pause certain aspects of your Cyber Risk Management Action Plan to prioritize the ongoing corporate transition.

Conducting these regular meetings also allows new information to emerge, which may necessitate adjustments to your Cyber Risk Management Action Plan. Despite thorough data gathering, unexpected discoveries may still occur during your plan's execution. For instance, you might realize your Protect function is weaker than initially anticipated. During these instances, it is important to reconnect with the original purpose that motivated you to embark on this cybersecurity journey in the first place. Always take the time to remind yourself of the underlying reasons for pursuing reasonable cybersecurity and what is truly at stake. By reconnecting with your purpose, navigating unexpected information and making informed decisions becomes easier.

In addition to monitoring progress and addressing challenges, these monthly meetings present an opportunity to celebrate the accomplishments and successes of your organization in the realm of cybersecurity. How you choose to celebrate reflects your company's culture, but its significance should not be underestimated. While it is essential to recognize achievements across all areas of the business, celebrating cybersecurity wins in a distinct way can inadvertently reinforce the notion that cyber risks are separate from other aspects of the business.

To overcome this counterproductive narrative, it is crucial to integrate cybersecurity celebrations seamlessly into your overall company culture. Treating cybersecurity milestones and achievements with the same level of enthusiasm and recognition as any other business accomplishment, you foster the perception that cybersecurity is an integral part of regular business operations. This approach helps establish a mindset where individuals perceive cybersecurity as a natural and essential component of their everyday professional responsibilities.

Celebrating cybersecurity wins alongside other business achievements reinforces the idea that effective cybersecurity practices are not isolated or detached from the organization's broader objectives. Instead, they become ingrained within the fabric of your company's regular business life. This alignment promotes a holistic and proactive approach to cybersecurity, where individuals recognize its importance and actively contribute to maintaining a secure and resilient digital environment.

During the monthly meetings, the final agenda item discusses the next steps in your cyber risk management journey. If a particular cybersecurity function has achieved its target score, you can mark it as completed and shift your focus to the next priority on your list.

For instance, if your organization's physical security is identified as a concern due to the absence of access control badges with individual photographs, your project may involve implementing a system to incorporate photographs on the organizational security badges. This project includes tasks such as procuring the necessary systems, organizing the badging and photography process, and establishing administrative procedures to ensure that every new member receives a badge with their photograph. Once the project is successfully handed off to the physical security team, and they have commenced its operation, the risk associated with inadequate physical security has been mitigated, marking the completion of that specific project. At this point, updating your cyber risk records to reflect progress and adjusting your priorities accordingly is essential.

Regularly reassessing and updating your cyber risk management initiatives ensures that your organization stays proactive in addressing vulnerabilities and implementing effective risk mitigation strategies. This iterative process enables you to tackle one project at a time, continuously improving your cybersecurity posture and safeguarding your organization against evolving threats.

STEP 3: SCHEDULE QUARTERLY REVIEWS

While the monthly meetings focus on tactical aspects, it is equally important to schedule quarterly reviews to assess the overall progress every 90 days. These reviews provide an opportunity to reflect on implementing various risk mitigations and evaluate the improvements made in cybersecurity scores during the previous quarter.

Managing and setting expectations regarding product progress, risk reduction, and the business value generated during the quarterly reviews is crucial. You should always seek to emphasize not only the reduction in risks, but also the tangible benefits achieved, such as increased productivity. Highlighting the positive business outcomes resulting from your cybersecurity efforts is essential during these meetings.

Another important agenda item for the quarterly review is to look beyond your organization and consider the broader cybersecurity landscape. In order to prevent becoming the next victim of a cyber attack, your organization must stay vigilant and informed about the evolving threats and changes in the overall cybersecurity landscape. This awareness may require adopting new products, providing training on emerging technologies, or updating processes to align with the evolving landscape.

Unlike the monthly meetings that focus on implementation goals, quarterly meetings should involve a wider range of participants. These meetings keep stakeholders informed about the progress and steer the overall Cyber Risk Management Action Plan. If your organization operates in highly regulated industries like banking, insurance, or healthcare, involve sales leaders to address questions related to cybersecurity that may arise during their interactions with potential customers. This ensures they are equipped with up-to-date information and can address any potential concerns their clients might have.

Another thing you should be prepared to do during your quarterly meetings is to answer any questions or concerns raised by the organization's key stakeholders. The perception and support of these stakeholders play a vital role in practicing reasonable and repeatable cybersecurity. By understanding and addressing anxieties or challenges raised, you can take proactive steps to reinforce confidence and mitigate potential obstacles, as well as use this opportunity to manage and maintain the reputation of your Cyber Risk Management Action Plan.

One of the most crucial aspects of the quarterly review is updating the scores from your original cyber risk scorecard. Based on the actual progress made, you need to revise the scores in your spreadsheet.

As your organization completes various risk mitigations, these improvements should be tracked, and a noticeable change in the organization's cybersecurity scores should be observed over time. Additionally, it is important to consider any new or evolving internal and external risks to the organization. If these threats warrant adjustments to the scores or an overall change to your priorities, they should be documented and added to your Cyber Risk Management Action Plan.

By conducting regular quarterly reviews, you can ensure ongoing monitoring, adaptation, and alignment of your cyber risk management efforts with the changing cybersecurity landscape. These reviews facilitate informed decision-making, enhanced risk awareness, and the ability to effectively respond to emerging challenges.

STEP 4: SCHEDULE AN ANNUAL CYBERSECURITY SUMMIT

Hosting an annual cybersecurity summit provides an opportunity to reflect on your company's progress throughout the year. It sets the stage for repeating the phases of the Cyber Risk Management Action Plan process as you get ready to move into your second year.

While similar in content to monthly and quarterly meetings, the summit has an even broader focus and serves as a comprehensive overview of your annual journey. During the summit, you should showcase your previous scorecard alongside the latest scores, emphasizing the improvements made in cybersecurity. This visual representation highlights the proactive measures taken by the organization to enhance its security.

This summit is a time for celebration, acknowledging the achievements made and recognizing the contributions of stakeholders who have made it all possible. Emphasize that as you enter a new year, you will face a new set of top cyber risks together. It is important to always demonstrate your commitment to ongoing cyber risk management as an organizational team.

It is time for you to reenergize the company as you embark on another round of the systematic, comprehensive, and structured process of practicing reasonable cyber hygiene. You will inform participants that we will start with a new interview series shortly. Just as Peter Drucker, a renowned management theorist, emphasizes, an effective executive must always set and adhere to priorities. Once the initial high-priority tasks have been accomplished, you reassess and determine your next set of priorities. This process applied to our annual interviews, as well, since they help us answer the question, "What actions are now necessary to uphold our reasonable cybersecurity efforts moving forward?"

Following the annual summit and the completion of this third phase of the CR-MAP process, you will return to the beginning and initiate the data-gathering phase again. This spiral and iterative process not only effectively manages cyber risks and adds value to your organization, but also enables you to construct a comprehensive narrative in the event of a data breach. This story, supported by comprehensive interviews, detailed meeting minutes, comprehensive scorecards, and other artifacts, demonstrates your proactive approach to cybersecurity and helps establish your company's diligence and preparedness. By leveraging these resources, you can mitigate potential consequences and avoid the fate suffered by other organizations and their customers that have been victimized by cyber-attacks and data breaches.

SUMMARY

In this chapter, we explored the importance of storytelling as you progressed through your cybersecurity journey and the Cyber Risk Management Action Plan process. Effective storytelling is crucial in garnering support and commitment from stakeholders within your organization. This third phase of your Cyber Risk Management Action Plan provides an opportunity to share a compelling story with key stakeholders, ensuring that they continue to buy-in to your greater vision for the organization.

Implementing cybersecurity measures requires people to change their daily operations, and it is vital to provide them with a story they can easily comprehend. Through the regular cadence of monthly, quarterly, and annual meetings, you have the platform to narrate the story of your collective journey. These gatherings serve as valuable occasions to engage stakeholders, communicate progress, and reinforce the significance of cybersecurity.

By crafting a compelling narrative and effectively sharing it during these meetings, you can foster a deeper understanding and appreciation for the importance of cybersecurity among your organization's members. This heightened awareness will contribute to their support and commitment to maintaining a secure environment.

Remember, your cybersecurity story is an ongoing process that evolves with time. Embrace the power of storytelling to create a shared vision, inspire action, and cultivate a culture of cybersecurity within your organization.

CHAPTER FIFTEEN

CONCLUSION

Congratulations on completing this comprehensive guide to mastering cyber resiliency and preparing for the AKYLADE Certified Cyber Resilience Fundamentals (A/CCRF) and AKYLADE Certified Cyber Resilience Practitioner (A/CCRP) certification exams. Throughout this book, we have provided you with the knowledge, skills, and practical insights necessary to navigate the complex world of cybersecurity and implement the NIST Cybersecurity Framework effectively within your organization.

We began our journey by establishing a strong foundation in the basics of the NIST Cybersecurity Framework. From understanding the core principles and functions to exploring the various categories and subcategories, you have gained a solid understanding of the framework's theoretical aspects. This knowledge will serve as the building blocks for your implementations within the world of cyber resiliency.

Then, we dove into the practical application of the NIST Cybersecurity Framework using the Cyber Risk Management Action Plan

(CR-MAP). By examining real-world scenarios, you have witnessed firsthand how organizations across different industries have successfully implemented the framework to enhance their cyber resilience. Armed with this practical knowledge, you are now equipped to be a cyber resiliency professional as you apply the framework effectively within your organization.

We have emphasized the importance of hands-on practice and continuous improvement throughout your learning journey. If you intend to take the certification exams, we highly recommend that you first complete the practice exams for the A/CCRF and A/CCRP certification exams, available to download for free at https://www.akylade.com/mastering-cyber-resilience. These practice exams are designed to provide you with an opportunity to test your knowledge and gauge your readiness for the official certification exams.

Once you complete the practice exam, you should review the correct answers and explanations included with the practice exams to further reinforce your understanding and address any knowledge gaps.

Remember, the pursuit of cyber resiliency is an ongoing endeavor. The field of cybersecurity is always evolving, and it requires a proactive and adaptive approach to stay ahead of the new and emerging threats that are discovered daily. By embracing the concepts and principles covered in this book, you have confidently acquired the tools to navigate the complex cybersecurity landscape.

As you embark on your certification exams, remain focused, trust in your preparation, and apply the knowledge you have gained throughout this journey. Passing the AKYLADE Certified Cyber Resilience Fundamentals (A/CCRF) and AKYLADE Certified Cyber Resilience Practitioner (A/CCRP) certification exams will validate your expertise and enhance your professional standing within the cybersecurity community.

Finally, we would like to express our gratitude for choosing this book as your guide. Remember that cybersecurity is a collective effort, and you play a vital role in safeguarding digital assets and protecting organizations from threats. Together, let us continue to foster a more cyber resilient future as we seek to ensure the security and resilience of our interconnected, digital world.

We hope that these certification exams are a steppingstone in your successful cybersecurity career and that your cyber resilience efforts ensure your organization's future.

APPENDIX A

A/CCRF EXAM OBJECTIVES

The AKYLADE Certified Cyber Resilience Fundamentals (CRF-002) exam consists of five domains:

Domain 1	Framework Concepts	25%
Domain 2	Framework Core	30%
Domain 3	Implementation Tiers	10%
Domain 4	Framework Profiles	15%
Domain 5	Risk Management	20%

Domain Objectives/Examples	Questions	Chapter
1.1 Identify key terms related to the NIST Cybersecurity Framework - Cybersecurity - Information security - Information systems security - Information assurance - Cyber resilience	2	2, 3

- Cybersecurity incident - Stakeholder - Supplier - Critical infrastructure - Threats - Vulnerabilities - Confidentiality - Integrity - Availability - Non-repudiation - Authentication		

1.2 Summarize key aspects of the NIST Cybersecurity Framework
 	
- Purpose of the NIST Cybersecurity Framework
- Components of the NIST Cybersecurity Framework
  - Framework core
  - Framework profiles
  - Implementation tiers
- Six functions of the NIST Cybersecurity Framework
  - Govern
  - Identify
  - Protect
  - Detect
  - Respond
  - Recover
	4	4, 5

1.3 Summarize how the NIST Cybersecurity Framework is different than other frameworks and certifications
 	
- Applicable sectors and industries
  - Government
  - Healthcare
  - Financial services
  - Energy
	3	4

- Manufacturing
- Retail
- Transportation
- Critical infrastructure
- Characteristics of the framework
  - Voluntary set of guidelines
  - Flexibility and adaptivity
  - Focus on risk instead of technical controls
  - Focus on risk instead of compliance requirements
  - Facilitate communication and collaboration
  - Continually improved and evolving
- Other frameworks and informative references
  - International Organization for Standardization (ISO)/International Electrotechnical Commission (IEC) 27001 and 27002
  - National Institute of Standards and Technology (NIST) Special Publications (SP 800-37, SP 800-53, SP 800-171, SP 800-218, and SP 800-221A)
  - Cyber Risk Institute (CRI) Profile
  - Center for Internet Security (CIS) Critical Security Controls
  - Information Technology Infrastructure Library (ITIL)
  - Payment Card Industry Data Security Standard (PCI-DSS)
  - Health Insurance Portability and Accountability Act (HIPAA)
  - North American Electric Reliability Corporation (NERC) Critical Infrastructure Protection (CIP) Standards
  - Federal Risk and Authorization Management Program (FedRAMP)		

- Open Web Application Security Project (OWASP) - Cloud Security Alliance (CSA) Security, Trust, Assurance, and Risk (STAR) Registry		

1.4 Explain the benefits of achieving cyber resilience to key stakeholders
 	
- Development of the NIST Cybersecurity Framework
- History of the NIST Cybersecurity Framework
  - Executive Order 13636
  - Executive Order 13800
  - Executive Order 14028
  - Cybersecurity Enhancement Act of 2014
  - Relevance of NIST Cybersecurity Framework to contemporary cyber risks
  - Federal Information Security Modernization Act (FISMA) of 2014
  - Cybersecurity Information Sharing Act (CISA) of 2015
	1	4

2.1 Explain the importance of the framework core - Purpose of the framework core - Usage of the framework core - Benefits of the framework core - Effectiveness of the framework core	2	6

2.2 Explain how categories are utilized within the six functions
 	
- Govern (GV)
  - Organizational Context (GV.OC)
  - Risk Management Strategy (GV.RM)
  - Roles, Responsibilities, and Authorities (GV.RR)
  - Policy (GV.PO)
	5	6

- Oversight (GV.OV)
- Cybersecurity Supply Chain Risk Management (GV.SC)
- Identify (ID)
  - Asset Management (ID.AM)
  - Risk Assessment (ID.RA)
  - Improvement (ID.IM)
- Protect (PR)
  - Identity Management, Authentication, and Access Control (PR.AA)
  - Awareness and Training (PR.AT)
  - Data Security (PR.DS)
  - Platform Security (PR.PS)
  - Technology Infrastructure Resilience (PR.IR)
- Detect (DE)
  - Continuous Monitoring (DE.CM)
  - Adverse Event Analysis (DE.AE)
- Respond (RS)
  - Incident management (RS.MA)
  - Incident Analysis (RS.AN)
  - Incident Response Reporting and Communication (RS.CO)
  - Incident Mitigation (RS.MI)
- Recover (RC)
  - Incident Recovery Plan Execution (RC.RP)
  - Incident Recovery Communication (RC.CO)		

2.3 Explain how subcategories are utilized with the six functions
 	
- Govern (GV)
  - Organizational Context (GV.OC)
  - Risk Management Strategy (GV.RM)
  - Roles, Responsibilities, and Authorities (GV.RR)
  - Policy (GV.PO)
  - Oversight (GV.OV)
	3	6

- Cybersecurity Supply Chain Risk Management (GV.SC)
- Identify (ID)
   - Asset Management (ID.AM)
   - Risk Assessment (ID.RA)
   - Improvement (ID.IM)
- Protect (PR)
   - Identity Management, Authentication, and Access Control (PR.AA)
   - Awareness and Training (PR.AT)
   - Data Security (PR.DS)
   - Platform Security (PR.PS)
   - Technology Infrastructure Resilience (PR.IR)
- Detect (DE)
   - Continuous Monitoring (DE.CM)
   - Adverse Event Analysis (DE.AE)
- Respond (RS)
   - Incident management (RS.MA)
   - Incident Analysis (RS.AN)
   - Incident Response Reporting and Communication (RS.CO)
   - Incident Mitigation (RS.MI)
- Recover (RC)
   - Incident Recovery Plan Execution (RC.RP)
   - Incident Recovery Communication (RC.CO)		

2.4 Summarize how the NIST Cybersecurity Framework outcomes are related to controls provided by other publications - International Organization for Standardization (ISO)/International Electrotechnical Commission (IEC) 27001 and 27002 - National Institute of Standards and Technology (NIST) Special Publications (SP 800-37, SP 800-53, SP 800-171, SP 800-218, and SP 800-221A)	2	7

- Cyber Risk Institute (CRI) Profile - Center for Internet Security (CIS) Critical Security Controls - Information Technology Infrastructure Library (ITIL) - Payment Card Industry Data Security Standard (PCI-DSS) - Health Insurance Portability and Accountability Act (HIPAA) - North American Electric Reliability Corporation (NERC) Critical Infrastructure Protection (CIP) Standards - Federal Risk and Authorization Management Program (FedRAMP) - Open Web Application Security Project (OWASP) - Cloud Security Alliance (CSA) Security, Trust, Assurance, and Risk Registry (STAR)		
3.1 Explain how implementation tiers are utilized in the NIST Cybersecurity Framework, including how they differ from a maturity model - NIST Cybersecurity implementation tiers - Capability Maturity Model Integration (CMMI) - Cybersecurity Capability Maturity Model (C2M2) - Cybersecurity Maturity Model Certification (CMMC) - ISO/IEC 27001	1	8

3.2 Given a scenario, analyze an organization's implementation tier based on its current cybersecurity posture - Tier 1 (Partial) - Tier 2 (Risk Informed) - Tier 3 (Repeatable) - Tier 4 (Adaptive)	2	8
3.3 Given a scenario, recommend strategies for moving an organization between implementation tiers - Assess the current state - Define the target state - Develop a plan of action - Implement the plan of action - Monitor and adjust	1	8

4.1 Summarize how profiles are used to tailor the Framework for varying risk management strategies
 	
- Key components of a profile
  - Core functions
  - Categories
  - Subcategories
- Utilizing profiles
- Current profile versus target profile
- Map profiles to an organization's cybersecurity posture
	3	9

4.2 Given a scenario, utilize a profile to tailor the NIST Cybersecurity Framework to specific organizational needs - Tailor profiles to support risk management strategies - Tailor profiles to support regulatory compliance requirements - Utilize profiles to measure an organization's cybersecurity posture over time - Identify relevant core functions, categories, and subcategories	2	9

4.3 Explain the use of profiles in the NIST Cybersecurity Framework
 	
- Profile templates
- Sector-specific profiles
  - Cyber Risk Institute (CRI) Profile
  - Manufacturing Profile
  - Election Infrastructure Profile
  - Satellite Networks Profile
  - Smart Grid Profile
  - Connected Vehicle Profiles
  - Payroll Profile
  - Maritime Profile
  - Communications Profile
	1	9

5.1 Explain the fundamentals of risk management
 	
- Risk Analysis
  - Qualitative
    - Likelihood of a risk
    - Impact of a risk
  - Quantitative
    - Single-loss expectancy (SLE)
    - Annualized loss expectancy (ALE)
    - Annualized rate of occurrence (ARO)
	2	3

- Hybrid
- Business Impact Analysis
  - Recovery time objective (RTO)
  - Recovery point objective (RPO)
  - Mean time to repair (MTTR)
  - Mean time between failures (MTBF)
  - Single point of failure
  - Mission essential functions
  - Identifying critical systems
- Financial Analysis
  - Total cost of ownership (TCO)
  - Return on investment (ROI)
  - Return on assets (ROA)
- Risk appetite
- Risk tolerance		

5.2 Given a scenario, determine the appropriate risk response to a given threat or vulnerability
 	
- Risk Responses
  - Acceptance
  - Avoidance
  - Transference
  - Mitigation
- Risk Register
- Types of Risk
  - Inherent risk
  - Residual risk
	2	3

5.3 Given a scenario, assess cybersecurity risk and recommend risk mitigations - Identify threats to an organization - Identify vulnerabilities to an organization - Identify risks to an organization - Recommend specific risk mitigations - Determine benefits of a particular risk mitigation	4	10

- Determine the trade-offs of a particular risk mitigation - Evaluate the effectiveness of a particular risk mitigation - Develop a risk management plan - Develop a cybersecurity strategy		

APPENDIX B

A/CCRP EXAM OBJECTIVES

The AKYLADE Certified Cyber Resilience Practitioner (CRP-002) exam consists of four domains:

Domain 1	CR-MAP Fundamentals	20%
Domain 2	Phase One: Determine Top Cyber Risks	36%
Domain 3	Phase Two: Creating a CR-MAP	27%
Domain 4	Phase Three: Maintenance and Updates	17%

Domain Objectives/Examples	Questions	Chapter
1.1 Explain how to best prepare for an assessment - Understand the target organization - Create project roadmap	1	11

1.2 Understand the CR-MAP process - Prepare needed documents - Contextualize the plan for the organization	1	11
1.3 Given a scenario, coordinate with management to achieve organizational buy-in - Provide adequate answers to management questions - Communicate potential business impacts of cybersecurity incidents - Communicate complex technical topics using layperson terminology - Create a communication plan to achieve buy-in	2	11
1.4 Explain the relationship between the NIST Cybersecurity Framework (CSF) and the Cyber Risk Management Action Plan (CR-MAP) - Understand how CR-MAP questions relate to the NIST Cybersecurity Framework outcomes - Understand the zero to ten scale used in CR-MAP as it related to the NIST CSF	1	11
1.5 Given a scenario, establish a risk profile for an organization - Understand details about a target organization to align them with a NIST CSF risk profile	1	11

2.1 Given a scenario, determine the appropriate stakeholders and create a list of interviewees to identify cyber risks - Consider role and technical ability - Consider geographic locations and branches	2	12
2.2 Given a scenario, conduct interviews, and record responses to identify top cyber risks - Present questions in an unbiased manner - Provide example answers to questions - Record interviewee notes	2	12
2.3 Given a scenario, analyze network diagrams to identify cyber risks - Review subnetting data - Review VLAN data - Review Virtual Private Network (VPN) data - Review legacy systems data	1	12
2.4 Given a scenario, assess any missing details after gathering data and remediate the missing details - Review qualitative data - Review quantitative data - Conduct additional interviews, as needed	2	12

2.5 Given a scenario, create and present the top cyber risks report for an organization - Display the top cyber risks by business unit in aggregate - Contextualize the top risks using themes - Generate high-level remediation recommendations for top cyber risks		1	12
2.6 Given a scenario, generate a custom questionnaire for an organization - Assign technical questions to interviewees - Assign non-technical questions to interviewees - Remove any non-applicable questions		1	12
2.7 Given a scenario, create charts to visually explain the top cyber risk categories to an organization - Create spider/radar diagrams - Create bar graphs - Create pie charts - Analyze raw data		1	12
2.8 Given a scenario, set an organization's target scores for alignment with the NIST Cybersecurity Framework - Understand the zero through ten CR-MAP scale in relation to the NIST Cybersecurity Framework		1	12
3.1 Given a scenario, verify how each top risk is covered by the mitigation roadmap		1	13

3.2 Given a scenario, rate each mitigation's business value based on the Business Value Model - Financial returns - Technical risk mitigation - Legal risk mitigation - Reliability of operations	1	13
3.3 Given a scenario, create custom mitigations based on organization questionnaire and interviews	2	13
3.4 Given a scenario, create standard operating procedures (SOPs) for custom mitigation and control - Implement mitigations - Implement controls	1	13
3.5 Given a scenario, generate a cost estimate for each mitigation and control - Understand the common cost associated with given mitigations and controls - Recommend contractors for mitigations and controls you cannot advise on	1	13

3.6 Given a scenario, create an implementation roadmap for an organization
 	
- Assign mitigations to specific organizational units
- Group mitigations by the type of owner
- Generate Gannt charts
- Understand resource limitations
  - Time
  - Money
  - Skilled personnel
	2	13

4.1 Given a scenario, assist leadership in assigning mitigations and controls to internal and external parties	1	14
4.2 Given a scenario, generate an updated top cyber risk presentations as mitigations and controls are implemented - Review completed mitigations - Determine scores assigned to each mitigation - Update charts with newly received numeric data - Update top cyber risks	2	14

4.3 Given a scenario, explain which mitigations and controls have been proposed and what they accomplish
 	
- Understand recommended mitigations/controls
  - Technical controls
  - Administrative controls
  - Physical controls
  - Preventative controls
  - Detective controls
  - Corrective controls
	1	14

4.4 Given a scenario, conduct ongoing reviews and maintenance of the organization's cyber resiliency - Create post-assessment communication plans - Update organizational roadmap based on periodic reviews	1	14

APPENDIX C

GLOSSARY

This glossary references all the terms used in the exam syllabus and the official textbook. These key terms and definitions should be understood by candidates prior to taking their certification exams.

3TCO (three-year total cost of ownership)
　　The comprehensive cost estimation of a solution or system over three years, incorporating both implementation costs and annual operating costs

annual operating cost
　　The total expenses incurred on an annual basis to maintain and operate a particular system, service, or solution, including renewal costs and ongoing labor expenses

annualized loss expectancy (ALE)
　　A metric used to estimate the expected financial loss over a specified time period resulting from a particular risk

annualized rate of occurrence (ARO)
A metric used to represent the estimated frequency at which a specific risk event is expected to occur within a year

Adverse Event Analysis (DE.AE)
A category of the Detect function that focuses on the analysis of anomalies, indicators of compromise, and other potentially adverse cyber events to characterize the events and detect cybersecurity incidents

assessment
An internal management activity focused on identifying areas for improvement

Asset Management (ID.AM)
A category of the Identify function that involves the identification of assets, data, hardware, software, systems, facilities, services, and people that enable the organization to achieve its business purposes

asset value
Represents the financial worth of the asset at risk

attorney-client privilege
A legal concept that protects communications between a client and their attorney from being disclosed to third parties to ensure confidentiality of information shared during legal advice or representation

audit
An external evaluation aimed at finding faults within the organization

authentication
The process of verifying the identity of individuals or entities attempting to access digital systems or resources to prevent unauthorized access and ensure data security

availability
The assurance that digital systems, services, and resources are accessible and usable when needed, without disruptions or services being denied

Awareness and Training (PR.AT)
> A category of the Protect function that emphasizes the importance of educating and raising awareness among personnel about cybersecurity risks, threats, and best practices to foster a security-conscious culture and enhance the organization's overall cybersecurity posture

big city approach
> A modern and mature perspective to cybersecurity in which the organization's Response and Recover functions are the priority instead of heavily focusing on the Govern, Identify, Protect, and Detect functions

business impact analysis (BIA)
> A risk management process used to examen the potential impacts of disruptions on an organization's systems, processes, and operations using a systematic evaluation to identify and prioritize critical systems and functions, assess their dependencies and interdependencies, and establish recovery objectives

category
> A group of related cybersecurity outcomes that collectively comprise and NIST Cybersecurity Framework function

Capability Maturity Model Integration (CMMI)
> A process improvement approach that provides organizations with a set of best practices to enhance their capabilities and achieve higher levels of maturity in software and systems development

CIANA pentagon
> The five core principles of cybersecurity (confidentiality, integrity, availability, non-repudiation, and authentication) that form the foundation for protecting digital assets and maintaining secure environments

community profile
> A baseline of NIST Cybersecurity Framework outcomes that is created and published to address shared interests and goals among a particular sector, subsector, technology, threat type, or other use case

compliance architecture
: The structure and framework that organizations establish to ensure adherence to regulatory and legal requirements related to cybersecurity and data privacy, which involves the design and implementation of policies, processes, controls, and technologies that enable the organization to meet its compliance obligations

confidentiality
: The protection of sensitive information from unauthorized access or disclosure by ensuring that only authorized individuals or entities can access and view confidential data

Continuous Monitoring (DE.CM)
: A category of the Detect function that ensures that the information system and assets are continually monitored to identify anomalies, indicators of compromise, and potentially adverse cybersecurity events

controls
: The specific measures, practices, or safeguards that organizations implement to manage and mitigate cybersecurity risks

core (also known as the framework core)
: A taxonomy of high-level cybersecurity outcomes that can help an organization manage and mitigate its cybersecurity risks through the use of functions, categories, and subcategories for each outcome

critical infrastructure
: Any physical or virtual infrastructure that is considered so vital to the United States that its incapacitation or destruction would have a debilitating effect on security, national economic security, national public health or safety, or any combination of these

critical system
: Any system whose failure or disruption would have a significant impact on the organization's ability to deliver essential services or fulfill its mission

current profile
　　A depiction of an organization's existing cybersecurity practices, including its cybersecurity activities, desired outcomes, and current risk management approaches

cyber resilience
　　An organization's ability to withstand and adapt to cyber threats by effectively responding to and recovering from cyber attacks or disruptions while minimizing damage and maintaining essential functions

Cyber Risk Institute (CRI) Profile
　　A strategic tool for financial institutions that provides a sector-specific roadmap for navigating the complex cyber risk landscape using the NIST Cybersecurity Framework version 2.0

cybersecurity
　　The practice of safeguarding computer systems, networks, and digital information from cyber threats through a range of technical, operational, and managerial measures aimed at preventing unauthorized access, attacks, data breaches, and other malicious activities in the digital realm

Cybersecurity Capability Maturity Model (C2M2)
　　A maturity model developed by the U.S. Department of Defense (DoD) to assess and improve the cybersecurity capabilities of defense contractors based on their maturity across various domains uses a three-level scale of Initiated, Performed, and Managed to assess organizational maturity

Cybersecurity Enhancement Act of 2014
　　A United States regulation signed into law in December 2014 aimed to strengthen and advance cybersecurity research and development efforts in the United States

cybersecurity incident
　　Any unauthorized or malicious event that compromises the confidentiality, integrity, or availability of an organization's digital assets, systems, or networks

Cybersecurity Information Sharing Act (CISA) of 2015
A United States regulation that facilitates the sharing of cybersecurity threat information between the government and the private sector

Cybersecurity Maturity Model Certification (CMMC)
A maturity model developed by the U.S. Department of Defense (DoD) to assess and certify the cybersecurity maturity of organizations participating in DoD contracts that consists of three levels, moving from level 1 (foundational cyber hygiene) to level 2 (advanced cyber hygiene) to level 3 (expert cyber hygiene)

Cybersecurity Supply Chain Risk Management (GV.SC)
A category of the Govern function that ensures the cyber supply chain risk management processes (an increasingly common vector for cyber attacks) are identified, established, managed, monitored, and improved by organizational stakeholders

Data Security (PR.DS)
A category of the Protect function that focuses on protecting the confidentiality, integrity, and availability of sensitive data within an organization's systems and networks, ensuring appropriate safeguards are in place to mitigate data breaches and unauthorized access

Detect (DE)
A function used by organizations to develop and implement appropriate activities to identify the occurrence of a cybersecurity event through timely discovery and analysis of anomalies, indicators of compromise, and other potentially adverse events

Executive Order 13636
The presidential executive order signed by Barack Obama in 2013 that aims to improve critical infrastructure cybersecurity by establishing a framework for information sharing and collaboration between the government and private sector entities

Executive Order 13800
The presidential executive order signed by Donald Trump in 2017 emphasizes the need for executive branch agencies to implement the NIST Cybersecurity Framework and encourages the private sector to

also adopt the framework to improve risk management and better prioritize cybersecurity investments across various sectors

Executive Order 14028

The presidential executive order signed by Joseph Biden in May 2021 that aims to strengthen the cybersecurity of federal networks and improve information sharing between the U.S. government and the private sector on cyber threats, incidents, and risks

exposure factor

The percentage of loss that would occur if the asset were compromised

Federal Information Security Modernization Act (FISMA) of 2014

A United States regulation that amended the FISMA of 2002 to emphasize the adoption of risk-based approaches and the use of industry standards, including the NIST Cybersecurity Framework, to enhance the security posture of federal agencies and improve the protection of federal information systems

financial analysis

A crucial aspect of risk management that focuses on assessing the financial implications and considerations associated with cybersecurity measures and investments

first responder approach

An approach that focuses the organization's resources on building out a fast, high-quality response capability in order to mitigate the other functional areas having lower target scores

function

The highest level of organization for cybersecurity outcomes in the NIST Cybersecurity Framework: Govern, Identify, Protect, Detect, Respond, and Recover

Govern (GV)

A function that involves creating, communicating, implementing, and monitoring an organization's cybersecurity risk management strategy, expectations, and policy

hybrid risk analysis
> A risk analysis approach that combines both the qualitative and quantitative approaches to assess risks, incorporating both subjective judgments and numerical metrics to gain a comprehensive understanding of the likelihood, impact, and financial implications of the identified risks

Identify (ID)
> A function that involves developing an organizational understanding of cybersecurity risks to an organization's assets, including its data, hardware, software, systems, facilities, services, people, suppliers, and capabilities

Identity Management, Authentication, and Access Control (PR.AA)
> A category of the Protect function that is used to implement effective mechanisms for the management of user identities, ensuring proper authentication processes, and controlling access to systems and resources to prevent unauthorized activities

impact
> The magnitude of a risk's consequences if the risk is realized

implementation cost
> The total expenses associated with acquiring and deploying a system, service, or solution, including acquisition costs and labor expenses for implementation

implementation example
> A concise, action-oriented, notional illustration of a way to help achieve a NIST Cybersecurity Framework core outcome

implementation tier (also known as framework implementation tier or the tier)
> An implementation tier represents the level of effectiveness in implementing cybersecurity practices within an organization, ranging from partial to adaptive

Improvement (ID.IM)

A category of the Identify function that is focused on improvements to organizational cybersecurity risk management processes, procedures, and activities that have been identified across all of the NIST Cybersecurity Framework Functions

Incident Analysis (RS.AN)

A category of the Respond function that ensures that proper analysis is conducted to respond effectively and to support the organization's forthcoming recovery activities

Incident Management (RS.MA)

A category of the Respond function that focuses on how an organization manages their response efforts for detected cybersecurity incidents

Incident Mitigation (RS.MI)

A category of the Respond function that ensures activities are performed to prevent the expansion of an event, mitigate its effects, and resolve the incident

Incident Recovery Communication (RC.CO)

A category of the Recover function that ensures all restoration activities are coordinated with internal and external parties, such as with their coordinating centers, Internet Service Providers, owners of attacking systems, victims, other cybersecurity incident response teams, and vendors, as appropriate

Incident Recovery Plan Execution (RC.RP)

A category of the Recover function that focuses on the execution and maintenance of the recovery processes and procedures to ensure than an organization's systems and services have been restored after a cybersecurity incident to full operational availability

Incident Response Reporting and Communication (RS.CO)

A category of the Respond function that focuses on ensuring that all response activities are coordinated with internal and external stakeholders as required by laws, regulations, or policies

information assurance

A comprehensive approach to managing and safeguarding information assets, encompassing technical controls, people, processes, and technology to ensure the confidentiality, integrity, availability, and non-repudiation of information, as well as the implementation of policies, procedures, training, and risk management frameworks

information security

The protection of information and data assets from unauthorized access, use, disclosure, alteration, or destruction that involves the implementation of security measures, policies, procedures, and controls to ensure the confidentiality, integrity, and availability of information

information systems security

The protection of computer systems and the associated infrastructure that store, process, transmit, and manage information and encompasses the security measures, policies, and controls implemented to safeguard computer hardware, software, networks, and databases from unauthorized access, attacks, and disruptions

informative references

A mapping that indicates a relationship between a NIST Cybersecurity Framework core outcome and an existing standard, guideline, regulation, or other content

inherent risk

The level of risk that exists in an organization's systems or processes without any control measures or risk mitigation efforts in place

integrity

Ensuring that data remains accurate, consistent, and unaltered throughout its lifecycle by protecting it against unauthorized modification, deletion, or corruption

ISO/IEC 27001 maturity model

A framework that assesses the maturity level of an organization's information security management system based on the ISO/IEC 27001 standard and provides a structured approach for organizations to evaluate their current state of information security practices and measures organizational progress toward achieving higher levels of maturity

likelihood
The probability of a risk event occurring or being realized

maturity model
A structured framework that assesses and guides the progression of an organization's capabilities and maturity levels in a specific domain, providing a roadmap for improvement and growth

mean time between failures (MTBF)
The average duration between two consecutive failures of a system or component

mean time to recover (MTTR)
The average time required to restore a failed system or process to full functionality after an incident

minimum score approach
An approach that sets out to achieve a minimum score across the board based on the belief that this is reasonable within the organization's industry, its customer expectations, and its organizational capability

mission essential functions (MEFs)
The key activities or processes that an organization must perform to maintain its core operations and fulfill its mission

National Institute of Standards and Technology (NIST)
A non-regulatory federal agency within the United States Department of Commerce whose mission is to promote innovation and industrial competitiveness by advancing measurement science, standards, and technology in various fields, including cybersecurity, manufacturing, energy, healthcare, and others

NIST Cybersecurity Framework (CSF)
A set of guidelines, best practices, and standards developed by the United States government based upon input by experts in the private

industry to help organizations manage and improve their cybersecurity risk management process

non-repudiation
The assurance that the originator of a digital communication or transaction can neither deny their involvement, nor the authenticity of the data being exchanged

Organizational Context (GV.OC)
A category of the Govern function that involves understanding the circumstances surrounding the organization's risk management decisions, including the mission, stakeholder expectations, dependencies, and legal, regulatory, and contractual requirements

organizational profile
A mechanism for describing an organization's current and/or target cybersecurity posture in terms of the NIST Cybersecurity Framework core's outcomes

outcome
The desired result or objective that an organization aims to achieve through implementing cybersecurity controls and practices, focusing on those measures' overall effectiveness and impact

Oversight (GV.OV)
A category of the Govern function that ensures the results of organization-wide cybersecurity risk management activities and performance are used to inform, improve, and adjust the risk management strategy

Policy (GV.PO)
A category of the Govern function that ensures that the organization's cybersecurity policy is established, communicated, and enforced

profile (also known as framework profile)
An organization's cybersecurity objectives, current state, and target state that provide a roadmap for aligning cybersecurity activities and priorities with the organization's business requirements

Protect (PR)
A function used by organizations to develop and implement safeguards to ensure the delivery of critical services and the protection physical and digital assets against cyber threats

Platform Security (PR.PS)
A category of the Protect function that focuses on the organization's consistent management of its risk strategy to protect its confidentiality, integrity, and availability of its hardware, software, and services of both its physical and virtual platforms

qualitative risk analysis
Risks are assessed based on subjective judgments, such as the likelihood and impact of a risk using a scale as opposed to using numerical metrics or figures

quantitative risk analysis
Risks are evaluated using numerical values and metrics to assess the financial impact and frequency of risk events

Recover (RC)
A function that helps an organization develop and implement appropriate activities to maintain resilience plans and restore any capabilities or services that were impaired due to a cybersecurity incident

recovery point objective (RPO)
The point in time to which data must be recovered following a disruption is defined by the maximum acceptable amount of data loss that an organization can tolerate

recovery time objective (RTO)
The targeted duration within which a business process or system must be restored after a disruption to avoid significant impacts, and it defines the maximum tolerable downtime for a specific process or system

residual risk
The level of risk that remains after implementing risk mitigation measures, such as controls and safeguards

Respond (RS)
A function used by organizations to develop and implement appropriate activities to perform actions regarding a detected cybersecurity incident

responsible disclosure
The ethical practice of promptly and transparently informing affected parties about discovered vulnerabilities or data breaches in order to mitigate potential harm

return on assets (ROA)
A financial ratio that measures the efficiency and profitability of an organization's use of its assets to generate earnings

return on investment (ROI)
A financial metric that assesses the profitability and financial benefits of an investment relative to its cost

risk
The potential for loss, damage, or harm resulting from the occurrence of threats exploiting vulnerabilities in digital systems or assets

risk acceptance
A risk response action that involves acknowledging the existence of a risk and choosing not to take further action to avoid, transfer, or mitigate it

risk appetite
An organization's willingness to accept potential risks related to its digital systems and assets, guiding decision-making processes to align risk management strategies with business objectives and priorities

Risk Assessment (ID.RA)
A category of the Identify function that ensures the organization understands the cybersecurity risk to its organizational operations (including mission, functions, image, or reputation), organizational assets, and individuals

risk avoidance
　　A risk response action that aims to eliminate or minimize risks by avoiding activities or situations that pose a significant threat

risk management
　　The systematic process of identifying, assessing, prioritizing, and mitigating potential risks to an organization's digital systems, networks, data, and assets to ensure their confidentiality, integrity, and availability

risk management lifecycle
　　A systematic and iterative approach to managing risks by encompassing several phases: risk identification, risk assessment, risk response planning, risk mitigation, and ongoing risk monitoring and review

Risk Management Strategy (GV.RM)
　　A category of the Govern function that ensures the organization's priorities, constraints, risk tolerance and appetite statements, and assumptions are established, communicated, and used to support operational risk decisions

risk mitigation
　　A risk response action that focuses on reducing the impact or likelihood of a risk event through the implementation of controls, safeguards, and countermeasures

risk register
　　A centralized document or database that systematically records and tracks identified risks, along with their attributes, assessment results, and corresponding risk management actions, to facilitate effective risk monitoring and mitigation

risk tolerance
　　The level of risk that an organization is willing to accept in pursuit of its objectives before action is deemed necessary to reduce it

risk transference
　　A risk response action that involves shifting the potential impact of a risk to a third party, typically through contracts, agreements, or insurance policies

Roles, Responsibilities, and Authorities (GV.RR)
: A category of the Govern function that establishes cybersecurity roles, responsibilities, and authorities to foster accountability, performance assessment, and continuous improvement are established and communicated

single loss expectancy (SLE)
: A metric used to estimate the potential financial loss that an organization may experience from a single occurrence of a risk event

single point of failure (SPOF)
: A component or resource that, if it fails, would cause a complete failure of an entire system or process

stakeholders
: An individual or group with an interest or influence in the organization's digital systems and assets, whose perspectives and requirements may shape risk management strategies and decisions

strong castle approach
: An approach that selects a target score profile that emphasizes the Protect function over the other five functions

subcategory
: A group of more specific outcomes of technical and management cybersecurity activities that comprise a NIST Cybersecurity Framework category

supplier
: An external entity that provides goods, services, or resources to an organization, and assessing the associated risks with suppliers is crucial to ensure they meet the organization's security and compliance requirements, minimizing potential vulnerabilities and threats introduced through their products or services

target profile
: The organization's desired state of cybersecurity practices and outcomes, as well as outlining the specific cybersecurity improvements and goals it aims to achieve

Technology Infrastructure Resilience (PR.IR)
> A category of the Protect function that ensures the security architectures are managed in alignment with the organization's risk strategy to protect the confidentiality, integrity, availability of its assets while maintaining its organization resilience

threat
> Any potential source or actor that has the capability to exploit a vulnerability, weakness, or flaw in order to cause harm to an organization's digital systems, networks, or data

total cost of ownership (TCO)
> The overall cost associated with owning, operating, and maintaining a particular asset or investment over its entire lifecycle

vulnerability
> A weakness or flaw in a system, network, or software that threat actors can exploit to compromise the security and integrity of digital assets

world-class approach
> An approach where every functional area is treated as equally important and a target score of 8 is assigned to all six functional areas

ABOUT THE AUTHORS

KIP BOYLE is the co-founder and Chief Operating Officer of AKYLADE, LLC and the founder and CEO of Cyber Risk Opportunities, whose mission is to help senior decision makers overcome cybersecurity hurdles. His customers have included the US Federal Reserve Bank, Boeing, Visa, Intuit, DuPont, Mitsubishi, and many others. A cybersecurity expert since 1992, he was previously the director of wide area network security for the Air Force's F-22 Raptor program and a senior consultant  for Stanford Research Institute (SRI). Kip has a graduate certificate in executive leadership and a master's in business management.

JASON DION is the co-founder and Chief Product Officer of AKYLADE, LLC, and founder of Dion Training Solutions, who strives to help candidates pass their cybersecurity, IT service management, and project management certifications. To date, he has helped over 1 million students across 195 countries get certified and advance in their careers. With decades of real-world experience, he has served as an Information Systems Officer, Director of a Network Operations and Security Center, the global lead for cyber defense for U.S. Cyber Command, and a Director of Information Assurance Operations (DIAO) for the National Security Agency, amongst other high profile cybersecurity positions. Jason holds a Master of Science degree in Information Technology with a specialization in Information Assurance (IA) and a Chief Information Officer (CIO) Graduate Certificate from National Defense University's College of Information and Cyberspace (CIC).

ALYSON LADERMAN is the Chief Executive Officer of AKYLADE, LLC. A Florida Trend Magazine "Florida Legal Elite," a 2024 Super Lawyer, and an AV-rated attorney by Martindale-Hubbell since 2009, Alyson has successfully handled complex and high-stakes cases involving contract disputes, business torts, intellectual property, employment, and insurance. She has been at the forefront of the merging of technology and law for more than two decades and has over a decade of executive leadership and C-Suite level experience.
Alyson earned her Juris Doctor cum laude degree from the University of Miami School of Law, and her Bachelor of Science with distinction from Nova Southeastern University.

Made in the USA
Columbia, SC
24 January 2025

52510687R00159